TIMEWARPS

Also by John Gribbin

WHITE HOLES:
Cosmic Gushers in the Universe

TIMEWARPS

John Gribbin

Delacorte Press / Eleanor Friede

Published by
Delacorte Press/Eleanor Friede
1 Dag Hammarskjold Plaza
New York, N.Y. 10017

This work was first published in
Great Britain by J. M. Dent & Sons Ltd.

Manufactured in the United States of America

Designed by MaryJane DiMassi

LIBRARY OF CONGRESS CATALOGING IN PUBLICATION DATA

Gribbin, John R
Timewarps.

Bibliography: p. 189
Includes index.
1. Time. I. Title.
QB209.G66   529   78-31165

ISBN 0-440-08509-8

To the memory of John W. Campbell, Jr.,
and for Harry Stead. Without them, I might
never have written this book—
or any other.

He was part of my dream of course—
but then I was part of his dream too!

LEWIS CARROLL
Through the Looking Glass

CONTENTS

INTRODUCTION

The concept of time is one of the most fundamental and at the same time one of the most baffling of our modern society. Our lives are ruled by the passage of time marked out on clocks, watches, and calendars; yet who doesn't feel, at some time or another, that there is more to the mystery of time than this orderly passage? The idea of time*warps* is, at least in a fictional sense, as basic a part of popular imagery as the idea of an orderly flow of time, thanks to science fiction in book, film, or television form; and all of us have wished to change time to give ourselves a second chance—the "if only" syndrome. Meanwhile, sober scientists with impeccable academic credentials and years of research experience calmly inform us that, in any case, time isn't something that flows inexorably forward at the steady pace indicated by our clocks and calendars, but that it can be warped and distorted in nature, with the end product being different depending on just where you are measuring it from. At the ultimate extreme, supercollapsed objects—black holes—can negate time altogether, making it stand still in their vicinity, while some physicists claim to have detected particles that actually travel backwards in time.

Where does this leave the average baffled observer of the modern scientific scene—or even the observer who has some training in physical sciences but still relies on clocks and calendars to run his everyday life? This is essentially the position I arrived at after experience both as a research astronomer and as a reporter and science writer covering the latest advances in physics and astronomy. Even while dealing with stories about the physical Universe—the "space" bit of what we should think of as a four-dimensional whole, "space-time"—aspects of the mysteries of time and timewarps kept creeping in. At the same time, an increasing—if perhaps belated—awareness of some of the universal mysteries that lie outside the realm of "science" directed my attention toward some similar mysteries from a different direction. The scientific mysteries of time and physical timewarps became, for me, a prelude to more philosophical ideas about the relationship between time and the human mind and the reality of phenomena such as precognition and reincarnation.

The structure of this book closely follows the development of my own awareness of these mysteries, from the everyday view of time through the intriguing concepts of physics and mathematics to the relationships between time and mind. In a sense, the whole book follows in this direction from my previous book on the physical mysteries of the Universe, *White Holes,* which could equally well have been titled *Spacewarps.* But although this structure may be a logical and orderly way to approach the subject of time and time travel, it helps to have a signpost indicating where we are going. So I was particularly pleased when, late in 1977, while working on the logically ordered structure you now see before you, I came across a very recent study that indicates exactly where the whole thing is going, and deals with one of the most intriguing aspects of the relationship between time and mind.

In an article published in the highly respectable and prestigious *Journal of Nervous and Mental Diseases,* and reported in some detail in more accessible form in the London *Sunday Times,*[1] Dr. Ian Stevenson presented evidence gathered from a 20-year study of 1,600 cases of reincarnation. Dr. Stevenson's work shows a clear standard pattern across these different cases and persuasive evidence for a real knowledge of "previous" lives among young children.

The typical pattern is for a child aged between 2 and 4 to suddenly start telling parents and others that he remembers a previous life, while at the same time showing a behavior pattern that fits well with the stories of a past life but may be strikingly different from that appropriate to his present surroundings. Such a child often asks to be taken

---

[1] September 11, 1977, "Children who have 'lived before,' " by Peter Watson.

to the place where, he claims, the former life was lived; in many cases this has resulted in identification of the family of the dead person who seems to have been "reincarnated," and (in Dr. Stevenson's words) "the child is then usually found to have been accurate in about 90 percent of the statements he has been making." But, after rising to a peak of obsession and detail between the ages of 3 and 5, typically the "memory" of a previous life then starts to fade and the children continue to develop along "normal" lines.

What triggers such behavior? Dr. Stevenson points to a high incidence of violent death in the previous personality that is "remembered"—sometimes suicide and often murder, certainly something that might make the previous life stand out from the average run of everyday human lives. He also claims that children in his sample often show phobias that echo this violence in the past life—a fear of water when death was by drowning, fear of automobiles when death was in a car accident, and so on. In many cases the children rejected their parents and family violently, saying they didn't belong; and a few ran away to seek their "real" parents.

All this is interpreted by Dr. Stevenson within the framework of reincarnation. To an outsider, however, it is not at all clear that the actual transfer of personality or a living "soul" from one human being to another is what is involved here. What *is* clear—and established to the satisfaction of the editors of the *Journal of Nervous and Mental Diseases,* who have both their own reputations and that of their journal to consider—is that some real phenomenon affecting the human mind is involved. Equally clearly, it seems to me, this phenomenon involves some warping of time. Perhaps, rather than being reincarnated, the children "view" past lives, coming into contact in some way with the personalities of people who have died in violent ways. How can such a thing occur? Astonishingly, such communication across the time barrier is quite within the bounds of possibility as outlined by current scientific knowledge and as discussed in Part Two of this book; and it is indeed appropriate to look at such a puzzle from the dual perspective of philosophy/psychology and the physical sciences. That is what this book is all about: a modest recognition by a trained physical scientist of the need to go outside physical science to get an understanding of the deepest mysteries of the Universe; an attempt to lead others along the same path that led me to this conclusion, and to show where the dual perspective might most fruitfully be applied; and, by no means least, a simple attempt to answer the age-old question "Is time travel possible?"

JOHN GRIBBIN
July 1978

PART ONE

Time Like
an Ever-Flowing
Stream

CHAPTER
1
Back to the Roots

The study of time was both the first science and the first organized religion of mankind. Our present-day society has developed, with some setbacks but never a complete break, from the society of Neolithic (New Stone Age) farmers that grew up in the millennia following the retreat of the last great Ice Age. For practical purposes, the end of that great freeze can be set at 10,000 years ago; and it is no coincidence that as the climate improved, so that conditions became clement enough to encourage the development of agriculture, the first stirrings of something we can identify as civilization also began to occur.

From about 7000 B.C.—certainly within a couple of thousand years of the retreat of the ice—men were living in villages and cultivating crops. To do this successfully, they must have had an understanding of one of the most basic rhythms of time that affect life here on Earth—the annual cycle of the seasons, which has been so important to farmers from the Neolithic right up to the present day. Even more remote ancestors of ours must have already been aware of the cycle of day and night; and to people without artificial lighting it would be deeply ingrained common knowledge that the change in proportion of daylight to darkness ties in with changes in the patterns of stars visible at different seasons, at different times of the year. With a large Moon to watch as it went through its phases every 29 days or so, plus the patterns of stars shifting with the seasons and the changing length of the night—and with no distractions from entertainment such as TV—every member of the Neolithic community must have had a feel for these rhythms of cosmic time virtually unmatched by members of our modern industrial society.

Only in the past couple of centuries have we cut ourselves off from this basic attunement with the rhythm of the seasons, by congregating in cities, producing bright artificial lights, and working away from the plants and animals that respond directly to these changing natural patterns. But the old established patterns and rhythms of time still color much of our thinking.

Consider the concept of reincarnation, still so firmly established in the minds of many people, and in organized religions, in one form or another. Surely this is related to the basic observation, over many human generations, that the whole pattern of life on Earth is one of rebirth as the seasons follow one another. Every winter is followed by a spring; every old moon is replaced by a new; every night gives way to a new dawn, as the Sun itself is reborn. So the practical need for a way of counting off the months and seasons, for a calendar and farmer's almanac, was reinforced in primitive society by a religious motivation to understand the rhythms of time—or, if not to understand them, at least to enumerate, record, and predict them.

Throughout the world we find great early civilizations, based upon the Neolithic farming revolution, that devoted great efforts to observing the heavens and building better calendars. Histories of science written from the European point of view generally cite the example of Egypt, a good one for many purposes, if only because the Egyptians left many records that have come down to us to be interpreted. But this book is not a history of science; and my aim in this chapter is simply to indicate the great antiquity of man's probings into the mysterious nature of time. As an example of the great importance of research into the rhythms of time in Neolithic communities, I have chosen the stone circles and other megalithic monuments of Western Europe and the British Isles. These monuments indicate that, in the words of Gerald Hawkins,[1] "there was in 2000 B.C. a dedicated, life-consuming preoccupation with the sky" in this part of the world; and the greatest product of this preoccupation is the astronomical computer at Stonehenge.

---

[1] *Stonehenge Decoded.*

Stonehenge and the Rhythms of Time

There is no longer any doubt that Stonehenge *is* an astronomical computer, and very little room to doubt that this is the purpose for which it was built. In his writings on the subject, the eminent astronomer Sir Fred Hoyle has said

> It is *not* a speculation to assert that *we* ourselves could use Stonehenge I to make eclipse predictions. . . . while this does not *prove* that stoneage man did in fact use Stonehenge I for making eclipse predictions, the measure of coincidence otherwise implied would be quite fantastic.[2]

Equally, there is today no longer any evidence to support the once cherished belief that any great works of the culture of Western Europe dating from 2000 B.C. or before must have been stimulated by ideas spreading from the eastern Mediterranean—or were even direct copies of Greek and Egyptian works. A revolution in the ability of archaeologists to date ancient sites has been brought about through the combination of radiocarbon techniques and tree-ring dating (dendrochronology). In essence, the residual radioactivity of the tiny proportion of carbon-14 in organic remains gives an indication of their age in terms of "radiocarbon years." Because this "radiocarbon calen-

---

[2] *From Stonehenge to Modern Cosmology.*

6

dar'' doesn't quite run at the same steady rate as the regular rhythm of the seasons, it must be calibrated; this is achieved using identical radiocarbon measurements of the age of slivers of wood from ancient trees—slivers that can also be dated precisely, and independently, using the seasonal rhythms that have laid down successive layers of growing wood as tree rings. This independent dating applies the calibration to the radiocarbon ''clock.'' The details, and some of the shattering implications for archaeology, have been described by Colin Renfrew, professor of archaeology at the University of Southampton in England, in his book *Before Civilization*. What matters here is that the new, unambiguous dating technique reveals that the great stone-using culture of Western Europe *preceded* that of the eastern Mediterranean. The megalithic tombs (from the Greek, meaning ''big stones'') were built hundreds of years before the pyramids; and the culture that created—as its equivalent to the giant atomic particle machines of our present culture—the great observatory/computer at Stonehenge owed nothing to the diffusion of ideas from the East.

The construction of Stonehenge actually spanned the period from about 2600 B.C. to 1700 B.C., with three distinct phases of building. The details of these are not important here, except to note that the *oldest* construction (Stonehenge I) was the most sophisticated in astronomical terms, and that the later addition of the impressive upright stones with their stone lintels seems to have been carried out by people who had lost some of the earlier knowledge of the use of the ''computer'' and were stressing the important alignments that were still understood—perhaps in an attempt to ensure that this knowledge too would not be forgotten with the passing generations.

Credit for drawing widespread attention to the possible use of Stonehenge as an astronomical instrument—a monitor of the rhythms of time—goes to Gerald Hawkins for his studies in the early and mid-1960s, using an electronic computer of the modern kind to check the significance of the many alignments of stones and other markers at this great monument on the Salisbury plain in southern England. Public imagination seems to have been fired by this pleasant juxtaposition, with the modern electronic machine being used to "prove" that these remains were also, in their fashion, a computing tool used by man; certainly Hawkins played up the role of the modern computer as an almost mystical modern artifact in his two books dealing with the subject. Ironically, however, this juxtaposition tells us more about the attitude of mid-1960s man to the godlike computer than it does about Stonehenge. In fact, no sophisticated aids are needed in interpreting the possibilities of Stonehenge, and there is a nice irony in this that is worth recounting.

When Fred Hoyle became interested in the problem, and in the possible designed use of Stonehenge, he lacked access to a modern computer of the kind used by Hawkins in his study. So, accepting at face value Hawkins's claim that such an instrument was essential to the investigation, he left the problem in spite of his interest and turned to other things. Only a couple of years later—during the wet, blank days on a walking holiday in Scotland—did he turn again to the puzzle with the aid of pencil and paper to discover:

> I found the need for a digital computer to be an illusion. It took a couple of days to set up the mathematical form of the problem, and then no more than a few hours to work through the arithmetic. Within quite trivial margins, I confirmed all of Hawkins's results.[3]

This is a very reassuring discovery, since while Neolithic man may well have had access to the brain-power equivalent of a Fred Hoyle, he certainly did not have access to a high-speed electronic digital computer.

It is perhaps not too surprising to find that the avenue at Stonehenge is aligned with the rising point of the midsummer Sun, so that on Midsummer Day—and no other—the Sun rises, as viewed from the center of the construction, exactly over the so-called "heelstone." This is both a very obvious feature of the annual rhythm of time, and one that can be marked with only a little ingenuity. Much more surprising is the discovery that one of the most simple features of the complex of stone circles and other markers at Stonehenge holds a very deep astronomical significance. This structure is a simple rectangle marked by "station stones," one at each of its corners. To understand the significance of this, we need to understand a little about the way the Moon moves across the sky, and how the lunar rhythms fit into those of Sun and Earth.

[3] *From Stonehenge to Modern Cosmology.*

9

While the Sun moves through a cycle in which it rises further north on the horizon each day from midwinter to midsummer, then reverses to rise (and set) a little further to the south each day until the next midwinter, the Moon's cycle of rising and setting does not fit exactly into an annual pattern. The Moon does follow the same kind of pattern of variation, rising more to the north on some days and more to the south on others, but the cycle only repeats, as viewed from a point on Earth, over a period of 18.61 years—just about 18⅔ years in round terms. The nearest whole number of years that can be fitted reasonably well to this lunar rhythm is 56 years, the time it takes for the cycle to repeat three times. That number too had significance for the builders of Stonehenge—but first let's look at the significance of that basic, simple rectangle.

The short sides of the rectangle point toward the direction of midsummer sunrise (looking northeast) and in the other direction (southwest) toward the direction of midwinter sunset. As I've already said, there's nothing too remarkable in that. The long sides of the rectangle, perpendicular to these significant solar directions, provide equivalent lunar markers, pointing toward the most southerly rising position and the most northerly setting position reached by the Moon during its 18.61-year cycle. On its own, that doesn't seem too remarkable either, although it does require observations over decades rather than years in order to get the alignments just right. But because of the geometry of the Earth/Moon/Sun system, it is possible to construct a rectangle that provides these markers for both Sun and Moon *only at the exact latitude of Stonehenge*!

Even 20 or 30 miles to the north or south, this doubly significant rectangular observing marker could not be built.

This has very deep implications for the present understanding of the Neolithic society of Western Europe, and is so clearly a deliberate construction that no one can doubt the deliberate intent in the minds of the builders. With many constructions that can reasonably be interpreted as astronomical observatories of the same period scattered around the British Isles and nearby coastal Europe, the picture emerges of a fairly closely knit society, with fairly good travel facilities (presumably by boat), which developed a very thorough understanding of the rhythms of time that are most important to farmers. Presumably, the construction of Stonehenge at such a uniquely interesting site, astronomically speaking, had deep religious and what we would now call "scientific" motives. The designers and builders knew already, from their other observations, of the remarkable combination of Sun and Moon alignments possible from the Salisbury plain, and doubtless regarded it therefore as a holy site unequalled in the known world.

Even so, the effort involved in the construction of this and other Stone Age astronomical observatories must have been great; by implication, the society that did the building must have been both stable and rich—the parallel with Egypt and the pyramids is obvious. Stable, since it took decades—even hundreds of years—to see the great projects through to completion. Rich, because in the first place the society could support a group of wise men who were able to study the astronomical alignments and plan their great observatories rather than spend their time toiling in the fields or hunting game, and secondly because of the ability of that society to feed (and presumably house and clothe) the teams of workmen engaged in building observatories, not in any productive labor. An equally obvious parallel, in more modern times, is with the rich, affluent society of the United States, which was able to send men to the Moon so recently. Stonehenge and the associated megalithic observatories are the remains of a Neolithic equivalent of the exploration of space. One difference is that they were intended to be used in the exploration of time, and in particular its rhythmic variations, and another difference is that the instruments remain in working order after 4,000 years. So much for any remaining view of our ancestors as dirty, hairy ape-men shivering in caves in the millennia before Christ. These were a wealthy, sophisticated people with a deep understanding of the mysteries of time and the structure of the Universe, a knowledge unequalled until the Renaissance, and (judging from the size of the artifacts) a wealth unequalled in relative terms until the twentieth century.

So much is evident from the unambiguous evidence of the careful choice of the Stonehenge site. Much more can be guessed at, in speculative fashion, from the astronomical interpretation of some of the more subtle features of the Stonehenge observatory. Here we are on slightly less certain ground. The archaeologists, in many cases, still cannot bring themselves to accept the subtleties that astronomers find in the structure of the Stonehenge computer; but the astronomers are happy that if *we* can use the computer for sophisticated prediction of eclipses today, it seems most unreasonable to imagine that the builders of the instrument could not use it with equal adroitness.

Most of these subtleties hinge around the number 56, which I've already mentioned as so important to the lunar cycles. By a remarkable coincidence (if it was a coincidence!) there is a ring of exactly 56 marker holes forming a circle around the stones of Stonehenge. Because of the way the Moon's path through space shifts relative to that of the Sun, the pattern of solar and lunar eclipses contains a rhythm of 18.61 years—which, again, can be seen in whole-number terms as a threefold rhythm repeating (very nearly) every 56 years. Hawkins suggested in the mid-1960s that this could solve the long-standing puzzle of this circle of 56 holes—the Aubrey Holes, named after a seventeenth-century scholar who noted their existence and other features of Stonehenge. The holes might simply be a counting device to record the passing of this triple cycle by moving a marker stone around the circle. It does seem, however, that you wouldn't really need a circle 320 feet across to keep count of a 56-year cycle in this way; and Fred Hoyle has developed the idea a stage further with some nice speculations in his version of the Stonehenge story.

Hoyle suggests that this circle marked out by the Aubrey Holes is what we would now call an "analog" computer—it can be used to track the movement of Sun and Moon—and in particular to predict the times when they are so close together that eclipses occur—by moving two markers around at the appropriate rates. A stone to represent the Sun must be moved two places every 13 days, taking 364 days to get round the circle. Not quite right for a year of just over 365 days, but Hoyle guesses that the position of the marker stone was reset (calibrated) at midsummer and midwinter, so that it was never more than half a day out. The Moon's orbit, on the other hand, takes 27.3 days, so its marker must be moved two spaces per day, starting from new moon or full moon when its position relative to the Sun marker is clear. This stone too needs resetting to keep errors from building up, and this can be easily carried out every new moon and full moon. In this model, the Aubrey circle represents what we now call the ecliptic—the plane in which the planets orbit around the Sun. Because the planets (and the Moon) wobble up and down relative to this plane, eclipses (which only occur when Sun, Earth, and Moon are exactly aligned) do not happen every month, but only at certain times in the 18.61-year lunar rhythm. These times can be predicted, using the same 56-hole circle, by using a third stone moved on by three holes each year. When *all three* stones are aligned appropriately, and only then, is there any chance of an eclipse.

This introduces a whole new standard of sophistication, and it is small wonder that many archaeologists cannot accept the implications. Now, according to Hoyle, we see the origin of a much more abstract concept. The third marker does not correspond to the visible gods of Sun and Moon, but to an abstract, theoretical point—an invisible god, and one moreover that is all-powerful, since this is the god that controls the eclipses of the other two. How did Stone Age man record the passage of thirteen days to keep his counters moving with the right rhythm? Simple, says Hoyle. With a circle divided into thirteen parts, and a stone to move one place each day, we can keep track by moving our main markers in the Aubrey circle each time our little circle (equivalent to a stopwatch, on the scale of the main Stonehenge circle!) has a stone in position 1 or position 7. Is it any coincidence, ponders Hoyle further, that we have a deeply rooted association of the number 13 with bad luck? Or that there are seven days in the week? Or even that one of our deep religious beliefs is in an almighty, invisible God as senior member of a Holy Trinity? The rhythms of time, it seems, are ingrained in our ancestral roots, our folklore, and even our religious beliefs. Going back to the very earliest times of which we have any records to interpret, the most remote ancestors of ours that we can ever hope to understand, we find this deep preoccupation with time going far beyond the basic needs of a simple farming community. Hoyle again:

> I suspect the mathematical methods of stoneage man are exemplified by this method of 7. I also suspect that much of our presentday culture—especially those things we grew up with and regard as so natural that we never bother to question them—we owe to our stoneage ancestors.[4]

In this book, my aim is to probe beyond those deep-rooted "commonsense" beliefs, to look more deeply at the mysterious nature of time. But in order to do so we need to de-

[4] See *From Stonehenge to Modern Cosmology* for a more complete account of the use of Stonehenge as an eclipse predictor.

termine the borders of common sense, and to make clear just what we really do understand by the "passage" of time—to acknowledge the place of time in the modern world. Having looked first at the roots of this common-sense view, back in prehistory at least as far *before* the time of Christ as we are now *after,* we can come up to date most conveniently by transferring our attention across the world. While the flowering of Neolithic culture in Europe passed—and even the rise of Mediterranean civilization, culminating in the Roman Empire, was followed by the long Dark Ages in which almost all scientific and philosophical continuity was lost—in one part of the world we trace the development of ideas almost from prehistory up to the present time. Chinese philosophers provide us with some of the oldest written thoughts about time and the nature of the Universe; and Chinese craftsmen produced the first timekeeping instruments that can be regarded as the direct forerunners of modern clocks. So on our travel forward in time from the Neolithic let's also move through space, from the Salisbury plain in Britain to the plains of the mystic East.

From Water Clock to Stopwatch: The Story So Far

Anyone who has received a typical "Western" education in recent years might have a vague recollection that the story of mechanical clocks begins in Europe in the fourteenth century A.D. with the invention of the mechanical escapement. More probably the only fact to have stuck in the mind will be the memorable definition of watches as one of the first consumer durables, which they became a couple of centuries after that invention, representing (together with cupboards) the kind of status symbols that are equivalent to pocket calculators and color TV in our modern society. And of course it is true that great progress in the accuracy of timekeeping instruments was made through the needs of sailors three or four centuries ago, since accurate observations of the heavens at precisely known times were essential for reliable navigation over long distances. But the story of timepieces actually goes back incredibly further still. Not in the fourteenth century A.D., not even in the seventh, but back within a couple of hundred years of the time of Christ, lie the roots of the development of modern timepieces; and it is no coincidence to find the ancient Chinese students of the rhythms of time as much concerned with the changing aspects of the heavens as a seventeenth-century navigator.

The story has been told by that remarkable interpreter of the history of Chinese science, Professor Joseph Needham, working with various colleagues.[5] These ancestors of our clocks were spheres, marked with a representation of the important stars, and driven steadily by water power to correspond to the apparent rotation of the heavens. In *Science and Civilisation in China,* Needham and Wang Ling quote a description of such an "armillary sphere" built around 132 A.D.:

[5] See especially *Heavenly Clockwork* and Volume 3 of *Science and Civilisation in China.*

> Chang Hêng made his bronze armillary sphere and set it up in a close chamber, where it rotated by the (force of) flowing water. Then, the order having been given for the doors to be shut, the observer in charge of it would call out to the watcher on the observatory platform, saying the sphere showed that such and such a star was just rising, or another star just culminating, or yet another star just setting. Everything was found to correspond (with the phenomena) like (the two halves of) a tally.

Other descriptions of Chang Hêng's work reveal that he built an armillary with two circles that showed the paths of the Sun, Moon, and five known planets as well as the "fixed" stars, all working indoors and compared with observations of the real heavens outside. And, says Needham, "nearly every succeeding century produced some astronomer or technician who accomplished the same thing." Yet surely the claim that the instruments agreed with observation "like the two halves of a tally" is something of an exaggeration, since these "clocks" driven by a water wheel must have had large inbuilt errors.

The principle of such devices was a water wheel with buckets around the rim, held from rotating by a trip device. Water pouring steadily into one bucket would increase its weight until the trip could be triggered, allowing the wheel to rotate one step and bringing the next empty bucket under the flow of water. This is a far cry from (though clearly a direct ancestor of) the accurate mechanical escapement at the heart of all modern clocks (up to the invention of "crystal" and "atomic" clocks). Such an early instrument must have deviated from the observed heavenly rhythms over a period of time; and another passage quoted by Needham and Wang Ling confirms this, providing at the same time a curious echo of the calibration technique that we suspect was used by the observers at Stonehenge. The passage describes the significance of observed differences between the mechanical "armillary sphere" and the real stars, the Sun, and the planets. The implication is that when observations of the sky were unobstructed by clouds, the mechanical device would be adjusted to keep in step with observations. Within its limitations, the sphere could then be used to predict the rhythms of the heavens, and especially the seasonal cycle. During cloudy nights and long rainy spells, the armillary sphere continued to rotate in as close an approximation of reality as possible—a tangible reassurance to the astronomers of the time that the cycle of the heavens continued uninterrupted even when it could not be observed; that the natural rhythms essential to life continued; that spring would follow winter; and that next year the crops would flourish as they had in the past.

This, in my view, is the great significance of such instruments. And they catered to the same basic human need as had the stone circles of northwestern Europe, two millennia earlier still and on the other side of the world. In Needham's words, "We do in a way foretell the future, says Su Sung, because we know that if the calendar is always well adjusted the work of the farmers will keep perfect time with the seasons and so . . . bring the best harvests."

Foretelling the future—there lies the key to man's obsession with time. First comes the realization that time passes (Where? What happens when it has gone?). Next we have the first understanding of the seemingly regular cycles of day and night and the seasons. Then calamity—the discovery that some cycles are not so regular. Eclipses don't always occur at the same time; each season is not an exact replica of the year before. And so we need reassurance. Will the seasons always follow roughly the same pattern, even if one year the frosts may be late in spring? Will the Sun always rise in the morning so that day follows night? And is there really an underlying rhythm to the apparently fickle fluctuations of those awesome phenomena, the eclipses? By pinning these rhythms down in mechanical devices—stone circles, water-powered armillary spheres, or whatever—we have reassurance and comfort. Yes, all is well. Day *must* always follow night; spring, even if late this year, will come along after the winter. Life will continue. Fulfillment of such a deep need goes far beyond anything we might call science or even religion. It is no exaggeration to say that the first, most important concern of self-aware man has always been his vital need to be sure that the rhythms of time will continue in a predictable pattern that makes it possible to plan ahead. Without such reassurance, there is no incentive to build for the future in any sense, and no prospect of any civilization ever arising.

So the search has continued for ever more accurate timepieces, to give ever more reassuring confirmation that we know what is happening and can plan ahead. From the

Chinese water-powered instruments of 19 centuries ago we can trace the development through the years, past such landmarks as the invention of the true escapement mechanism in about 723 A.D. (six centuries before European technicians reached the same stage) and on to the time when the cultures of East and West began to exchange information through interaction with Jesuit scientist-priests around the turn of the sixteenth/seventeenth centuries. It is tempting to speculate that the rather sudden appearance of the mechanical escapement in Europe in the fourteenth century, and the great speed with which the invention spread, may have marked an earlier transfer of information from East to West—but I must leave the investigation of such possibilities to those far more qualified than I to comment on them.

From the seventeenth century onward there is no question that the great developments in timekeeping precision originated in the burgeoning scientific civilization of Western Europe. By the nineteenth century, we begin to see the precision of timekeeping for scientific purposes—divorced from any practical activity, even navigation—approaching modern standards and simultaneously moving outside the province of everyday life. Astronomers—physical scientists—were moved inexorably toward consideration of the nature of time and its relation to space, an area of study previously the province of philosophers alone. The influx of ideas from physical science in turn stimulated the philosophers to ponder further on the nature of time and the Universe. While the poor man in the street was harried along, willy-nilly, into a twentieth century where travel on our small planet, the Earth, has become so fast that the natural rhythms of day and night become distorted; where scientists talk happily of stretching and compressing time; where even the ordinary TV-viewer must revise his understanding of "simultaneous" events and "instant" communication while listening to the pause between question and answer from Earth to lunar astronauts, made necessary by the finite time it takes signals to reach the Moon and return. The roots of our craving to understand time and be comforted by its reliable predictability may lie far in the past; but the confusion of modern understanding of the nature of "time"—at many different levels, intellectual and mundane—lies here and now. Our final need, then—before delving into the philosophical mysteries of time and the implications of relativity theory and eventually finding an answer to that key question "Can we travel in time?"—is to take stock of where we stand today. What does time mean in everyday terms in the last quarter of the twentieth century? How much of our "commonsense" view of what is going on around us is reality, and how much illusion? And most important of all, where do we go from here?

CHAPTER

Life on a Space-
Age Spinning Top

From an original obsession with the vital rhythms of time
expressed in the seasons and monthly cycles, we have now
"progressed"—if that is the right term—to a stage where
the ordinary citizen in the developed world is likely to
possess a timepiece with a hand to count off the seconds.
In sports, a precision of a hundredth of a second is commonly used to decide track records; in technology, the
precision, speed, and accuracy of our electronic computers
is becoming limited not through any failure of mechanical
precision, but by the tiny but finite time it takes for electric
pulses, traveling at the speed of light, to move along a
wire. Moving outward from this split-second precision,
time plays a vital and peculiar part in the life of the traveler, who is likely to suffer "jet lag" by moving around

the world between time zones. We can even jump from one day to the next, or repeat the previous day, by crossing the International Date Line—a kind of imitation "time travel" that has enabled some travelers to enjoy the benefits of two Christmas Days in succession, while others have had no Christmas at all in a particular year.

Movement of the Earth itself is now monitored so accurately that the basic astronomical rhythm of day and night is no longer regarded as a sufficiently precise "clock" with which to define our basic unit of time, the second. In the space age, the spinning top that is the Earth is regarded as a rather poor timekeeper, and we mark the passage of time not by the rhythm of the seasons or even by the cycle of day and night, but with mechanical clocks and messages from our local deejay. The midwinter festivals at Christmas and New Year are no longer linked with fundamental observations showing that the shortest day has passed, but are associated with numbers on a calendar; these European midwinter festivities are even celebrated today on the other side of the world, at the height of the southern summer! Small wonder, then, that few people today feel any direct affinity with the natural rhythms of time, or comprehend the way our mechanical substitutes are derived from and related to the real thing. With even navigators today as likely to steer by the aid of signals from an artificial Earth satellite as by the stars, it is left to astronomers, and their modern counterparts in the space sciences, to keep tabs on the spinning top Earth. So it is in astronomical terms that we can best develop an understanding of what the rhythms of time mean today: What are time zones and the Date Line? How can our biological rhythms be upset by travel? And so on.

More Days than One

The first complication we encounter, using studies no more difficult than the observations of our Neolithic ancestors, is that there are two ways to define a day. The more obvious is to call the interval between two successive passages of the Sun across the highest point in the sky one day, the interval between two successive noons. The alternative—easier in some ways, since the Sun is rather bright for direct observation—is to define the day in terms of the apparent movement of the stars across the sky, so that a day is the interval between successive passages of a chosen star across its highest point in the sky. But these two definitions—the solar day and the sidereal day—do not correspond, because the Earth itself is orbiting through space around the Sun.

All the stars are so far away that the movement of the Earth in one day round its orbit makes no difference, and on successive nights our chosen star will always be seen in the same direction in space. But in one day the Earth moves through a slice of its orbit approximately equivalent to one degree of arc (there are 360 degrees in a circle, and 365 days in a year, more or less). The result is that to observers on the spinning top Earth, it seems that the Sun has shifted by an equivalent amount against the background of the fixed stars. Between "noon" on one day and "noon"

25

on the next, the Earth must rotate not by just 360 degrees on its axis, but by roughly 361 degrees. As a further complication, because the Earth's orbit around the Sun is not an exact circle but slightly elliptical, the length of this solar day itself varies with the seasons, while the sidereal day is much more nearly constant. To astronomers, this is sufficient reason to base measurement of time on the sidereal day. But to farmers and anyone else concerned with the daily rhythm of life on planet Earth, it is much more natural to start from the basics of the solar day, even if these basics are not entirely unchanging.

But which solar day shall we use for our base? It's all very well to follow tradition and say that we'll have 24 hours in the day, 60 minutes in each hour, and 60 seconds in each minute—but we've got to choose one day to start from! In fact, astronomers have chosen as our base not one particular solar day, but one particular *average* of the changing solar day. The mean solar day is the average of all the solar days in any one year. But because the Earth is gradually slowing down in its orbit around the Sun, even this is not constant; and the basis of our measurements of time was established until recently in terms of the length of the year beginning January 1, 1900. Because the regular variations since then can be worked out precisely, it is always possible to relate time measurements back to this constant base. But now we have progressed even further, going off the astronomical standard to define time in terms of the rhythms not of the Sun and the stars, but of atoms and crystals.

Atoms and molecules emit and absorb electromagnetic

radiation (light or radio waves, for example) at certain well-defined frequencies that are related to the structure of the atoms and molecules. This provides chemists with an invaluable tool for identifying the presence of particular atoms in mixtures of substances, since each has its characteristic spectroscopic "signature." But since the frequency of the radiation is simply measured as a well-determined number of cycles per second, we can also turn the measurements around to provide a way to measure time.

Studies carried out in 1958 showed that the characteristic radiation associated with the element cesium had a frequency of 9,192,631,770 cycles per second, with an accuracy of plus or minus 20 cycles per second. The second used in this study was the standard second defined in astronomical terms, compared against the length of day over the year 1900. Since 1972, this remarkably accurate timepiece—the cesium atom—has formed the basis, of a new standard of time, Atomic Time (AT), in which one second is defined by the frequency relationship cited above. The basic cesium standards are used to calibrate slightly less accurate timepieces that keep time by monitoring the oscillations of quartz crystals. These timepieces are easy to operate (unlike the cesium standard) and can be run continuously—indeed, it is now commonplace to find such crystals used even in cheap watches instead of the more traditional clockwork.

So our basic standard of time today, which provides the time signals we get over the radio and use to calibrate our own clocks and watches, is derived from the atom, not from the stars. Because the poor old Earth—wobbling

27

about the Sun, sometimes spinning a fraction of a second faster and sometimes a fraction slower—is such an erratic timekeeper by these standards, we have occasional "leap seconds" introduced into the atomic standard to make sure that no one's watch need get more than one second out of step with the current mean solar day. A far cry indeed from the days when every effort of intellect, ingenuity, and the engineering skill was needed to keep accurate observation of the months, the seasons, and eclipse cycles— or even from the much more recent days when the introduction of the leap *year* cycle enabled mankind to establish a calendar that kept closely in step with the seasons, even though the length of the year is not exactly 365 days, but close to 365¼ days. But our atomic standard of time is, compared with astronomical time, as arbitrary and unvarying a standard as peeling off dates from a calendar. Perhaps we can best understand the relationship between our unvarying standards, be they atoms or calendars, by looking at leap years—which we have all experienced, if not fully understood—rather than by getting too deeply embroiled in the hairsplitting, second-dividing subtleties of Atomic Time.

The Whys and Wherefores of Leap Years

Just as with the day, it is possible to define the year either in terms of the apparent movement of the Sun or in terms of changes in the observed positions of the stars. Again, of course, it makes most sense to us to use the solar year as a basis, since otherwise we would find the seasons slowly drifting through our calendar as the years went by, with winter gradually slipping back into October, then September, August, and so on. There is nothing wrong with such a system, but it is more convenient to be certain that if someone reports events occurring in January in Europe, we can be sure that it was midwinter then, and don't have to work out the year first in order to find which season was passing through January at that time.

So we choose as our calendrical base the solar (or tropical) year, and try to keep the seasons as fixed as possible in terms of the annual calendar. That, however, is not entirely straightforward with a year that runs at present to 365 days, 5 hours, 48 minutes, and 46 seconds! Somewhere, somehow, we have to introduce fractions of a day into the calendar—and that is where leap years come in.

The first reasonably successful and widespread method of fitting a calendar with a whole number of days in each year to an astronomical system with an extra 5 hours 48 minutes and 46 seconds to be accounted for was developed two millennia ago, in the time of Julius Caesar. This calendar was proposed by the astronomer Sosigenes of Alexandria and ran on a four-year cycle in which three years of 365 days were followed by a fourth containing 366. This averages out to a year length of 365¼ days, or 365 days 6 hours—within 12 minutes of the true mean length of the tropical year. This was known as the Julian calendar, after Julius Caesar; and the rule for choosing leap years was the simple one that any year divisible exactly by 4 should contain the extra day, February 29.

But those eleven minutes or so of error continued to mount up as the years went by in this Julian system; and by 1583 the discrepancy amounted to a very noticeable ten days—the calendar had gone out of step with the heavens, with a consequent shift in the seasons. Clearly, something had to be done; and the man who did the necessary something (prompted by the astronomer Clavius) was Pope Gregory XIII. Introduced first in the then Catholic countries, the new rule, which gives us the Gregorian calendar, is that every year that divides exactly by 4 is to be a leap year *except* for century years (1800, 1900, and so on). Century years will be leap years only if they divide exactly by 400—thus, the year 2000 will be a leap year.

This makes things a lot more satisfactory. On average, over each 400-year cycle of repetition, the Gregorian year is within 26 seconds of the actual tropical year, so that it will take 3323 years from 1583 before the Gregorian calendar is even one day out of step with the heavens. The obvious way to correct this tiny error might be to introduce a new kind of leap year every 4,000 years or so—but the details of this refinement to the Gregorian calendar can safely be left for our descendants in the year 4906 to worry about, assuming we have any descendants who are still worried by such problems then!

Although it is the calendar we use today, it took some time for the Gregorian reform to spread around the civilized world. In England, and its colonies in America, there was no reform of the calendar until 1752, by which time the discrepancy with the heavens had reached 11 days. So, when the calendar was revised, it was necessary to introduce an Act of Parliament to the effect that the day after September 2, 1752, would be September 14, 1752. Even in the eighteenth century the primitive, superstitious nature of man's fundamental need to be reassured of the constancy of the rhythms of time expressed itself in widespread fear among the common people that this action had "robbed" them of 11 days of their lives. At the time, the calendar reform was blamed for many of society's ills, from bad health to bad weather and poor crops. Many people alive at the time were left in some confusion about their birthdays and when to celebrate them, a confusion that continues to the present day in the case of George Washington, born on February 11, 1732, by the old Julian calendar—by our modern Gregorian calendar, his anniversary is now celebrated on February 22.

In Russia, the reform of the Gregorian calendar wasn't introduced until 1918, after the Revolution. This has left us with the equally curious oddity that although the turning point in the Bolshevik seizure of power is now celebrated on November 7, the Revolution is still remembered in some quarters as the "October" Revolution—which indeed it was by the old calendar then in force in that part of the world.

So in fact it has taken us from the time of Neolithic man right up until the twentieth century to establish on a worldwide basis a calendar that provides us with a reasonable means of keeping our human records in step with the seasons. Why do the seasons themselves vary in a regular rhythm as the spinning top Earth rolls in its orbit around the Sun? Basically, because the Earth is tilted at an angle

31

to its orbit, so that first one hemisphere points toward the Sun and receives the benefit of summer heat, and then the other hemisphere receives the benefit of long, hot summer days while the other half of the Earth has cold, long nights and the frosts of winter. The details can be found in standard astronomy books, such as the excellent *College Physical Science* by Vaden Miles and colleagues. But these effects, and other subtleties in the wobbling, rolling movement of the Earth that may bring the rhythms of Ice Ages upon us,[1] are of no direct concern in our investigation of the puzzle of time. Far more significant to us now is the remarkable way in which, just seventy years after the whole world came over to the Gregorian calendar, even the common man and woman must come to grips with the subtleties of horological hairsplitting if they venture aboard a modern jet airliner and hurtle across the time zones.

Keeping Tabs on Time Zones

Because of the spin of the Earth, the Sun seems to rise every day in the East and set each night in the West. To understand and keep track of the time zone system, we can forget about the "seems to" and look at the situation from the very reasonable standpoint of a citizen of the Earth, pretending that the Earth itself is still. It takes only a little thought to see that if the Sun is rising in the East as we see

[1] See my book *What's Wrong with Our Weather?*

it (wherever we may be) then people to the east of us see it rise a little earlier than we do, and people to the west see it rise a little later. The same applies to the time of noon, when the Sun reaches its highest point in the sky; to sunset; and to all other times of day measured in terms of the Sun's movement across the sky. The wave of day and night sweeps around the planet in a never-ending rhythm, and every longitude on the Earth's surface must, in principle, have its own astronomical time. (The same applies, of course, if we choose the other means of astronomical timekeeping and monitor the nightly movements of the stars.)

It would clearly be ridiculous to take this time difference to hairsplitting lengths and have clocks in the eastern suburbs of a city set slightly ahead of clocks in the western suburbs. Just as we need to force the annual cycle into a calendar system with a whole number of days in each year, we need to force the different times at different parts of the Earth into a reasonably regular pattern in which the same time is used over a wide enough range of longitude that we don't have to reset our watches between leaving home and arriving at work, or for any comparable short journey. Obviously, we need to have clocks and watches set to the same time in each country or in each state. But how do we decide just where, and by how much, to make changes in our timekeeping when we do travel across state boundaries?

It turns out that there is a good natural way to break up the continuously changing rhythm of natural time into regular time zones, a way which works much more easily and naturally than the leap year system of our calendars. Because the Earth rotates through a circle (360°) every day (24 hours) the rotation in one hour is simply 15° (360 ÷ 24). So, measuring by the Sun, when it is 12:00 noon for us it will be 1 P.M. for people 15° of longitude to the east, and still only 11 A.M. for people 15° to the west. For every 15° further east, the time is one hour later still, and for every 15° further west, the time is one hour earlier still. (You may well ask what happens where East meets West 180° away on the other side of the world. Don't worry, there's an answer to that coming up!)

Now, one hour is a reasonable size for any change in our watches and clocks; it would be ridiculous to reset our timepieces by four minutes every time we traveled across 1° of longitude, but to change by one hour after a journey of 15° (further than the distance from New York to Detroit, or from London to Rome) is no great inconvenience, and can even add spice to the adventure of long-distance travel. All we need is to define a base longitude from which to start defining our time zones, and the rest of the world falls neatly into 15°-wide slices, like slices of melon.

It's not quite that simple, human nature being what it is, and geographic and national boundaries not always falling conveniently along the edges of our chosen time zones; but the system, if not always logical, is at least comprehensible. The natural starting point for the whole system was laid down in the days when Britannia ruled the waves and her geographers waved the rules. The meridian of longitude passing through Greenwich, the site of the old Royal Observatory, is defined as the 0° meridian, and the mean solar time at Greenwich (Greenwich Mean Time, GMT, or Greenwich Standard Time, GST) is the basis from which other terrestrial times are determined.

This basic time is also known as Universal Time (UT) and used by astronomers around the world regardless of local time. (Indeed, the nerve center of a modern observatory generally boasts an impressive array of clocks, at least one each for UT, AT, local civil time, sidereal time, and solar time; to the confusion of more than one visitor, these are not always labeled.) The 15°-wide time zones, one hour apart on the clock, are each centered on the corresponding meridian, so that GST, for example, should strictly apply over the belt of longitudes from 7½° west to 7½° east, and so on. But this is where the complications of human nature and geography come in. Neighboring states, especially those with economic and political links, often operate on the same time even if this means distorting the basic boundaries of the 15° natural time zones. In addition, so-called daylight saving time is used in many parts of the world to make maximum use of daylight hours. Under this system, clocks are set forward one hour in the spring, and back again in the fall. Working hours, established by the clock and not by daylight, are thereby shifted relative to the pattern of day and night, and our sleep patterns are adjusted so that we seem to gain an hour of daylight. But, as any farmer with livestock to care for will tell you, this kind of clock fiddling makes no difference to the natural rhythms of time. Cows that need milking an hour after dawn still need milking an hour after dawn, even if the clock now says 6 A.M. instead of 5 A.M.!

So, if you start out from GST and the 0° meridian, and then check the longitude of your home town and find out what the time ought to be, don't be too surprised if you find an hour, or even two hours, difference between your calculation and the time shown on all the clocks in town. But don't worry about it; as long as we know where the time zone boundaries are, and as long as they add up properly to give us 24 hours in a complete circle of the Earth, the system will work well whatever the local anomalies. The big crunch, though, as hinted above, comes at the 180° meridian.

Flying from east to west around the globe, you would cross all 24 of the 15°-wide time zones, and set your watch back by 24 hours, "losing" a day in the process. Flying eastward, the opposite would happen and you would "gain" a day. The paradox this raises is well demonstrated by the neat example of a jet flying westward at the same "speed" as the apparent passage of the noonday Sun, at latitude 42°N—a speed of only 700 miles per hour, quite feasible by present standards.[2]

At such a speed, travelers in the aircraft will see an unchanging Sun; if they leave on their journey at noon, it will still be noon when they return to their starting point after circling the globe. But noon on which day? To friends on the ground, clearly the "next" day, since they had a good night's sleep while the plane was aloft. But to the travelers? They have experienced no night at all, just a 24-hour-long "noon"! So somewhere on their journey the

[2] I have taken this example from the excellent description of the change-of-date situation in *College Physical Science*.

36

travelers have to make the change of day, quite arbitrarily, to remain in step with their earthbound friends. The most logical place to do this is on the opposite side of the world from the 0° meridian, at ± 180° longitude. Happily indeed for the geographers (who may well see in this a sign that the geographers of Britannia's heyday were playing their role as dividers of the globe not blindly but through the workings of some greater destiny) the 180° meridian runs almost entirely through the vast, sparsely populated wastes of the Pacific Ocean. To avoid the embarrassment of having not just a different time but a different day in neighboring regions of some communities, the International Date Line (IDL) zigzags away from the 180° meridian in a few places—around the tip of Siberia and around the Fiji Islands, for example—but otherwise runs pretty much due north-south across the ocean. On one side of the line, time is 12 hours ahead of Greenwich; on the other, 12 hours behind. So the difference on crossing the line is just 24 hours—one day on the calendar. Travelers crossing the IDL must change their watches not just by one hour but by a full day. From east to west, the traveler leaps a day forward—from midnight on December 24 to midnight on December 25, perhaps. The traveler moving eastward, however, gets the dubious benefit of "repeating" a day, and might jump back from 1 A.M. on January 1 to 1 A.M. on December 31, getting a chance to experience New Year's Eve twice. This is not really time travel, but simply a result of the conventional pattern of time zones and the arbitrary allocation of an International Date Line. The system works happily to keep intrepid circumterrestrial travelers on the same calendar as their more staid counterparts back home; and even astronauts, whizzing about the globe in a matter of a few hours, are in principle kept in step by the infallible IDL system—although in practice, of course, they keep the same time as their ground control station through their mission.

The most memorable beneficiary of the date line system is, alas, fictional—the hero of Jules Verne's *Around the World in Eighty Days*. In that story, the urgency of the frantic scurrying around the globe was provided by a bet that the journey could be achieved within 80 days; the hero accomplished his feat only to be thrown into jail (through an unfortunate mistake) on his return to Britain; he languished there overnight before the error was rectified. It seemed that he had lost his bet after all, since 81 days had now elapsed; but it turned out that, since he had been traveling from west to east, only 80 days had elapsed by the calendar back home. The day he gained crossing the IDL came to the rescue and enabled him to win his bet within the strict letter (but not, I feel, the spirit) of its original terms. Thus Verne provides us with a happy ending— avoiding the complexities that might have arisen on a similar east-to-west trip, when the happy traveler might well have arrived to claim his bet after 80 days only to find that 81 calendar days had elapsed. The moral here for all gamblers: Check the way the "track" is run, as well as the rules of the race, before placing any bets.

Confused? If, even in fiction, whirling about the globe in 80 days can make you dizzy, how about the problems of modern travelers on high-speed jets? Scarcely anyone would travel around the globe without stopping, but many people cross the Atlantic or Pacific ocean by jet, for business or pleasure. A goodly fraction of them suffer from disorientation in adjusting to the time of the zone in which they have landed—and this disorientation is not just a result of mental confusion. Like a great many—perhaps all—living organisms, the human body has its own inbuilt "biological clock"; and if you shift that body in a couple of hours from a time zone where breakfast has just been served to one where supper is ready—or one where another breakfast is due—the whole bodily rhythm can be thrown out of gear. Here, perhaps, lies one of the most important keys to the basic nature of time and its rhythms, as experienced by the inhabitants of one whirling planet.

Biological Clocks—A Basis of Behavior

It should be no surprise to find that present life forms on Earth, having evolved on the surface of a spinning planet for millions of years, should be well attuned to the natural rhythms of time associated with such a habitat. Some resulting patterns of behavior are indeed obvious: Some animals sleep at night and hunt by day, while others are nocturnal; the human menstrual cycle is strikingly similar to the length of the lunar cycle; the seasonal cycles are deeply ingrained in the reproductive patterns of both animals and plants. But the rhythms of cosmic time penetrate much deeper into the patterns of life on Earth—so deep, in fact, that it is only recently that the links between life and these cosmic rhythms have begun to be interpreted, and they are still far from being well understood.

Much of this work is still contentious. In the scientific establishment, there is often a reluctance to accept anything except overwhelming evidence that extraterrestrial influences can affect life on Earth—the whole thing smacks uncomfortably of astrology, and astrology has long been a dirty word in science. This is unfortunate. For, without in any way suggesting that the cheap kind of popular astrology typified by the horoscopes in the morning papers has any sound basis for determining the reader's way of life, it is now becoming clear to any unbiased observer—even one with a training in astronomy, such as myself—that there are very definite "outside" influences that affect both human life and other life on Earth.

Many of these links are physical and can be understood now that we have instrumented probes in space to monitor the changing interplanetary medium. Our Sun itself undergoes both regular and less regular variations, including a roughly 11-year-long cycle of activity, and sends blasts of particles across space in a "solar wind" that affects the atmosphere and weather of the Earth and disturbs the nearby magnetic field. This is perhaps the easiest outside influence to understand—if changes in the Sun affect the weather, there is no doubt that life is going to be affected by the changed weather patterns, whatever other direct effects on life may also occur. The exact nature of these physical links and their changing influence over past centuries is beyond the scope of the present discussion,[3] but it is surely worth a slight diversion to summarize some of the evidence for the rhythms of life, and their links with the rhythms of time, that has been gathered elsewhere in highly readable form by such excellent authors as Michel Gauquelin (*The Cosmic Clocks*) and Lyall Watson (*Supernature*).

An often repeated but still significant tale concerns the habits of oysters transported from their normal tidal homes to tanks—without tides—in a laboratory in the middle of the United States. In their natural habitat (in this case, Long Island Sound) the oysters attune their rhythm of life to the local tides, opening their shells at high tide and closing at low tide. When moved to tanks in Evanston, Illinois, the oysters at first retained the rhythm of home, opening up at the time of high tide on Long Island Sound—in much the same way that the human body experiences jet lag when skipping across time zones. But, just as humans adjust to local rhythms of day and night, the oysters soon adapted to life in their new quarters. By a couple of weeks after the move, the rhythm of opening and closing had shifted noticeably, until the oysters were in tune with a completely nonexistent tide—the sea tide that would have existed if the Illinois laboratory had been covered by ocean!

There is no doubt that this is a "tidal" effect—that the oysters can respond to the distant tug of the solar and lunar gravity fields even without the ocean tides to encourage

---

[3] See *What's Wrong with Our Weather?*

them. Covering the tanks to keep the oysters in the dark made no difference to their new rhythm, which kept in phase with the Moon, the oysters opening up when the Moon was overhead. And if oysters can respond directly to the passage of the Moon across the sky, why shouldn't the much larger human body, with its sensitive nervous system?

Among the wealth of evidence of a similar kind gained from many studies around the world the most interesting (at least to my human readers) must be that concerning people. All of us have an internal clock that controls our bodily functions in line with the daily rhythm—but the clock can be fooled, and gets out of step with the 24-hour rhythm unless it is constantly recalibrated against the pattern of day and night. This natural clock thus echoes the continual double-checking against the sky and recalibration of our mechanical clocks, from Stonehenge and the early Chinese water clocks right up to the regular adjustment of our present calendar through the leap year system. These daily rhythms of life are called *circadian* rhythms (from the Latin *circa,* about, and *dies,* day). The human circadian clock controls our body temperature, which changes during the day; it also controls the time of birth in humans, which peaks in the small hours, late at night before the dawn. And it is itself controlled not so much by the pattern of day and night as by any regular cycle or rhythm roughly 24 hours in length.

Just as workers on the night shift can adjust to a nocturnal existence, so people forced into a slightly longer or shorter day adapt to a new rhythm of time. How can you change the length of "day" on Earth? Ingeniously, one group of experimenters took volunteer subjects off to a region where there is no natural rhythm of day and night—the Spitzbergen Islands above the Arctic Circle, during the high summer period when the Sun never sets. Two groups of volunteers were established, one with its timepieces running on a 21-hour cycle and the other with a 27-hour day. And both soon adjusted to the new cycle, as shown by the change in body temperature and other circadian rhythms.

Some of the deepest-seated rhythms of life, however, could not be fooled. Just as the oysters knew where the Moon was even without an ocean tide washing over them, so the basic metabolic processes continue to follow the true 24-hour cycle even when sleep patterns and gross bodily functions such as temperature are changed. One such infallible circadian rhythm is the excretion from the body of potassium, used up in the nervous system as part of the mechanism by which signals are shunted around the body.

So the body has two kinds of daily rhythm, one easily fooled and the other more closely attuned to cosmic rhythms of time. In addition, there is the monthly rhythm already referred to (which also affects births, as the circadian rhythm does, with most births occurring around full moon and least around new moon); and perhaps, although this is more debatable, there is at least one kind of *annual* rhythm in people, which produces more births in the spring (May and June in the Northern Hemisphere). Studies of these rhythms, which continue in many places today, cover a whole spectrum of reliability and, depending on your basic prejudices, acceptability.

At one end are the undeniable, and acceptably scientific, studies of daily rhythms such as those of the Illinois oysters. At the other end, when we see evidence that time of birth is affected by the phases of the Moon, we tread dangerously close to the forbidden territory of astrology. Perhaps this is the moment to return to the mainstream of our present concern with the mysteries of time itself—mysteries that are in many ways the prerogative of the philosopher and the science fiction writer, and so unashamedly speculative that they offend no one. *Can* we travel in time? Even with the basic understanding of time, the physical time and its rhythms that affect our daily lives, that I have outlined in these two introductory chapters, the conventional answer to the question—an unequivocal "No"—remains founded solely in abstract philosophy. Time travel, we are told, is impossible because it leads to paradoxes. But those paradoxes may be more apparent than real—bizarre deviations from the normality of life on a spinning planet, but perhaps unremarkable in the broader framework of space, time, and the Universe.

CHAPTER

③

Travel in Time: Paradoxes and Possibilities

In all of the discussion so far—if you like, the "commonsense" view of time based on and relevant to our daily lives—the underlying thread has been the image of time flowing steadily forward. But this commonsense view is very much based on the here and now—the view appropriate to twentieth-century mankind in the "developed" world. As recently as a couple of centuries ago, the rigid view of time as a steadily flowing stream, inexorably carrying us along as our lives are ticked away by a multitude of clocks and watches, would have been inappropriate even to ordinary people in Europe, the most "advanced" part of the world at that time. Without a fixed schedule and cheap timepieces, the laborer in the fields took his lunch break when he felt hungry or when work permitted, not when the appointed hour for eating arrived. The whole pattern of human activity in such circumstances is governed more by subjective needs and bodily feelings, and less by the rigid application of outside schedules and allegedly objective timetables. Time may have flowed inexorably forward, always in the same direction, but to the peasant with no timepiece other than the Sun it certainly did not flow steadily. With more work to be done in the fields at some seasons, and with the proportion of day to night varying through the year, subjective time passed more rapidly some days—or some weeks and months—than it did at others.

Of course, we "know" that "really" each hour is the same length, and every day contains 24 of these hours, and time flows steadily as well as inexorably. But this "knowledge" is to some extent a result of our way of life. Subjective time is just as real to people—more real, in a sense. Philosophers have long debated the relation between subjective and objective time; more recently, psychologists have puzzled over the same problem; and in the world of physics and mathematics, where cool objectivity supposedly reigns supreme, scientists now happily accept the notion of stretching and compressing the flow of time, as we shall see in Part Two of this book. In such circumstances, it is not as big a step as it might seem (at least in philosophical, abstract terms) to consider just what the implications of real time travel would be.

Cycles of Time

I say "real" time travel here to exclude the obvious fact that, even within our modern philosophical framework, we all travel through time, at a rate of 24 hours a day, continuously. This rather boring progress at a constant "speed" and in a fixed direction (toward the future) is so commonplace that it is clear that when we talk about time travel what we mean is any deviation from this steady progress. Can we move faster forward, arriving at the future long before our friends who remain plodding along at 24 hours per day? Can we slow down our own passage through time, to be left behind by our more regular friends? Or, the most exciting possibility of all, can we travel at will forwards *or backwards* through time, like Dr. Who or the users of television's fictional Time Tunnel?

The practical possibilities, physical and mental, will be the subject of the rest of this book. But to set the scene, and to provide the bridge between the everyday experiences of time discussed in the first two chapters and the bizarre possibilities inherent in abandoning this commonsense view, it seems appropriate to look at some of the puzzles and paradoxes inherent in such "real" time travel—the paradoxes that, according to the unimaginative, are a sign that such time travel must be impossible. To the imaginative, however, the paradoxes are an exciting stimulus to further speculation and investigation. Maybe the world is not quite as it seems to our ordinary senses, going through the ritual of everyday life; and if it is not, what breathtaking visions open up before us! In the words of the fictional "Bylaw of Time" in Robert Heinlein's tale "All You Zombies," "A Paradox may be Paradoctored"—and that's where the fun comes in.

More of Heinlein—and other science fiction "philosophers"—shortly. But first let's look at some of the ideas about time held quite seriously by human societies in the not-too-distant past, and by some groups today. Pride of place must go to the Mayas of Central America, if only because they had the distinction of developing a calendar rather more accurate than the Gregorian version we use today. While our calendar produces an error of about 3 days in 10,000 years (unless we introduce some kind of leap millennium), the Mayan version lost only 2 days every 10,000 years (which makes it 5 days per 10,000 years different from the Gregorian calendar, which is "wrong" by three days in the opposite sense).

The Mayas, as G. J. Whitrow put it in his book *The Nature of Time,* were "of all people known to us . . . the most obsessed with the idea of time." But their obsession led to a quite different concept from our view of a steady, unidirectional forward-flowing river. Rather, the Mayan priests' "world view" was of a cyclic rhythm of time, containing lesser cyclic rhythms, so that history was thought to repeat over a rhythm of 260 years. Not necessarily an *exact* repetition was expected, but a general ordering of daily life in line with the trends of this dominant rhythm—so that, for example, when the European conquerors imposed their Christian religion on the Mayan society, this was seen as the echo of a previous religious up-heaval centuries before.

Other philosophies—or religions, for the names mean much the same in this context—have held more rigidly that events repeat themselves *exactly* in some regular cyclic rhythm of time. Whitrow quotes the writings of Nemesius, a fourth-century Bishop of Emesa, describing the beliefs of the Greek Stoics:

> The Stoics say that when the planets return, at certain fixed periods of time, to the same relative positions which they had at the beginning, when the cosmos was first constituted, this produces the conflagration and destruction of everything which exists. Then again the cosmos is restored anew in a precisely similar arrangement as before. The stars again move in their orbits, each performing its revolution in the former period, without variation. . . . Socrates and Plato and each individual man will live again, with the same friends and fellow-citizens. They will go through the same experiences and the same activities. Every city and village and field will be restored, just as it was. And this restoration of the universe takes place not only once, but over and over again—indeed, to all eternity without end.[1]

This concept finds a curious echo in one modern version of cosmology, which sees a pattern of birth in a cosmic fireball followed by phases of expansion, collapse back to the fireball, and then a repetition of the cycle.[2] And for the

[1] Quoted from Whitrow's *The Nature of Time*.
[2] See *White Holes*.

49

ancients—certainly for the majority of philosophers before the spread of Christianity—the variations on the theme of a cyclic Universe held much greater sway than the isolated support for the "linear" theory of a steadily progressing time flow. In such a Universe, there is no need for time travel—just wait long enough, and everything comes around again! But even so, perhaps, there might be some incentive to jump across the more boring parts of the loop of time, if that were possible. Even through the Middle Ages the philosophy of cyclic time remained prominent, though often parallel with the concept of linear time; and only since the time of Immanuel Kant—the middle part of the eighteenth century—can our modern idea of a steadily evolving, forward-flowing Universe be said to have held the center of the stage. Our "commonsense" view of time has been "obvious" for only a couple of centuries out of four millennia or so of the recorded study of time, going back to the society that built Stonehenge. With this sobering reminder that our common sense may be regarded as naive folly by future generations, we can perhaps look again at the implications of time travel unblinkered by twentieth-century conventions. We needn't go so far as to embrace wholeheartedly the Eastern philosophy typified by the Buddhist view that nothing—past, future, or physical space—has any "reality" except as an illusive form of thought.[3] But we can indulge in some Western philosophizing and speculation that is a little more comfortable, if still far removed from the "reality" of the daily routine.

[3] See Fritjof Capra's *The Tao of Physics*.

The Science Fiction Philosophers

The trouble with much of the philosophizing that is presented under the label of philosophy is that it might just as well be written in a foreign language for all its intelligibility to people who are not "philosophers." Since when you do dig through the verbiage to find some nugget of an idea, it often turns out that the idea is actually quite simple to understand, I sometimes wonder whether the rest is not just window dressing, camouflage to frighten off the nonexpert. Take the example of causality and the arrow of time—perhaps the single most important concept in the whole discussion of time travel, and the rock upon which the common belief that time travel is impossible is founded. In simple terms, the argument is just that an event must always occur *after* the event which caused it (hence the term causality). A bullet leaves the gun after the trigger is pulled; the results of a horse race reach us after the race is run and not before; and so on. This is common sense; we never experience events that violate causality in our everyday lives (we do not, alas, obtain the results of a race in time to place our bets on a certain winner). But does that necessarily mean that any implication that time travel violates causality must mean that time travel is impossible—the *reductio ad absurdum* argument? Most philosophers argue that it does; I feel that the case is unproven, and that, in addition, there may be aspects of time travel that do not violate causality at all.

51

There's no denying, however, that the question of causality is very much a central issue in the debate. But how do we get the debate into everyday language? Here's an example of the tortuous complexities philosophers can embroil themselves in when trying, quite laudably, to say precisely what they mean in unambiguous terms:

> It will be recalled that the A-Theory held temporal becoming to be an objective property of all events and claimed that because of this the past and future differ ontologically, the future being open and the past closed. Since past events have become present they have already won their ontological diplomas, unlike future events which still exist in a limbo of mere possibility.[4]

I don't entirely go along with the "A-Theory" anyway, as will become clear later in this book. But it is even more important to discuss just what is or is not believable in easily understood terms. Unfortunately, everyday language is inevitably colored by everyday experiences and everyday common sense—which, in fairness, is why the philosophers do not always use everyday language. Nevertheless, there is a body of literature that uses everyday language to put across quite complex philosophical arguments about time travel, causality, and the arrow of time; by drawing on that source rather than on weighty philosophical tomes, I hope to be able to look at the paradoxes and possibilities of time travel in intelligible everyday terms and without

[4] Richard M. Gale, *The Language of Time*.

unacceptable ambiguity. Bear in mind that if you want to delve into the real nitty-gritty of the implications, you'll just have to face up to the complexities of philosophical jargon. But meanwhile let's turn to those street philosophers, the science fiction writers—and readers.

The SF field is rich with speculation about time travel, causality violation, and all the trimmings—not always highly informed speculation, but generally entertaining and, where the cream is concerned, as valid as any speculations of the philosophers. In some cases, such as the writings of Sir Fred Hoyle, genuine philosophy is wrapped up in an entertaining SF package; and that, in many ways, gives us the best insight of all. So, with a last bow, for the time being, to the philosophers, let's take a look at the arrow of time and the implications of causality violation, bearing in mind that

> to bring out the conceptual centrality of the analytic truth that a present cause cannot have a past effect, we shall try to devise a counter-stipulation-example to this truth and shall note what corresponding conceptual reforms it requires us to make.[5]

Branches and Loops: The Two Basic Puzzles

In SF terms, the hoariest such example is of a time traveler who goes back in time and, wittingly or unwittingly, kills the young boy who would otherwise have grown up to be the killer's grandfather. So the killer was never born;

[5] Ibid.

he could never have gone back in time. In which case, he never killed his ancestor, who lived to a ripe old age, and in due course the killer was born. . . .

So the pattern repeats—or does not, as the case may be. To the everyday view of the world, the commonsense interpretation of causality, the existence of such a paradox in the hypothetical example of time travel is supposed to indicate that the concept is impossible. We might argue on such grounds that just as nature was once said to abhor a vacuum, she now abhors time travel. In SF literature, though, the paradox is merely the jumping-off point for further speculation. If the would-be killer can only exist if the killing fails to take place then something must intervene to sidetrack him; or it may turn out that the supposed grandfather is not, for one reason or another, the killer's grandfather at all. The most intriguing resolution of this classic causality violation "paradox" is the idea that everything that *might* happen in such circumstances *does* happen. The grandfather is both killed and not killed; the grandson both exists and does not exist. Just how this curious state of affairs can come about we shall be able to see after looking, in Part Two, at the implications of modern physics—ideas even more bizarre than the bulk of science fiction, and ripe with potential means of resolving dilemmas far more tricky than any simple time travel paradox.

The simplest resolution of the "paradox" is the idea that the effects of the time traveler's activities are already rooted in the fabric of time and space—that the effects follow causes in the time sense of the traveler, producing a fixed pattern of historical events. In Michael Moorcock's *Behold the Man* the time traveler is a disturbed individual, masochistic and with leanings toward religious mania, who journeys back to the time of Jesus in order to witness first-hand the events leading up to the Crucifixion. But his time machine is destroyed on arrival beyond repair—and there is no trace of Jesus in the form recorded in the Bible. Inexorably, the traveler is drawn into the Jesus role, taking over and playing out the pattern of events that he remembered from biblical tales of the "real" Jesus. So the fabric of history is preserved, and the Bible stories are told—and two thousand years later a certain disturbed, masochistic individual will go back in time to witness them.

Here, there is no obvious paradox to resolve. But we have a puzzle of a different kind, a closed loop in time which causes itself, the same kind of disturbing image as the snake that eats its own tail. Where now is the beginning, and where the end? What is the cause, and what the effect? No causality, or too much, is as great a problem in philosophical terms as broken causality! The classic example of the alternative puzzle to that posed by Moorcock in *Behold the Man* is Ward Moore's *Bring the Jubilee*—though I must admit that the book itself doesn't seem sufficiently inspired to me to merit its elevation to the status of an SF classic. In this variation on the time travel theme, the hero lives in a "parallel" world to our own—one that has the same history up until the American Civil War, but in which the Confederate States won that war.

Nothing terribly original there; parallel or alternate universes are a common SF theme, to which we shall return.

Moore's book seems to have gained its "classic" label through his depiction of the kind of society that might have developed after the North lost the Civil War; its merits are a matter of personal preference that I leave to the judgment of other readers. The link with our look at the puzzles and paradoxes of time travel comes from events that "follow" when the hero travels in time to—you guessed it—study firsthand the events of a critical battle in the war. Inadvertently he sets in motion a train of events that results in the North, not the South, winning that battle, and ultimately the whole war. The world he is left stranded in is quite different from the world of his remembered history books and the world he grew up in. Indeed, it is "our" world. Where now is the reality? What has happened to the "original" world? Did it ever exist? Or does it still exist, on its parallel time track, but cut off from our world?

The same theme is developed by L. Sprague de Camp in *Lest Darkness Fall*, but *starting out* from "our" time track, with a hero who is mysteriously deposited in sixth-century Italy and singlehandedly averts the Dark Ages. The tale is basically pure hokum and fun, with no pretensions of grandeur; but the author slips in his piece of philosophizing to explain how the hero can still exist even though history has been changed: Although the "web of history" (what a modern physicist might call "the fabric of space-time") is very tough overall, there are weak points where slippage between the centuries sometimes occurs. Metaphors become somewhat mixed with the further explanation that the "new" history resulting from the hero's activities in the sixth century (and this one really is a hero—at least he achieves heroic ends!) grows off from the main trunk of history as a new branch. Both the old history ("our" space-time) and the new exist growing from the same trunk, but quite independently of each other after the point of division.

Of course, you want to know how such a thing can happen physically, and that's where the author stops. After all, the tale is one of fiction (in fact, as originally published in the magazine *Unknown,* it was unashamedly labeled fantasy); but fiction, fantasy, and speculation are all quite allowable in philosophy, and all we are doing here is preparing the way for some bizarre physical ideas with a look at some philosophical implications of time travel.

And some people, who at least claim—perhaps tongue in cheek—that they are offering "serious" speculation, point to events in our own "real" history that look suspiciously like the activities of a time-transported hero diverting and disturbing the world line of our passage through space-time. Look at the diverse talents of Leonardo da Vinci, whose breadth of interests was as remarkable as the depth of his knowledge and talent—and whose designs for heavier-than-air flying machines would probably have worked had he had access to a modest internal combustion engine. Or take the much quoted example [6] of the maps discovered in the East in the nineteenth century by the Turkish naval officer Piri Reis. These maps, dating back over many centuries, show *all* the continents—including Antarctica—with great accuracy. Even allowing for hoax

[6] For instance, in Louis Pauwels and Jacques Bergier's *The Morning of the Magicians.*

or fraud at Piri Reis's time, it seems that in detail the maps are better than any that existed in the nineteenth century—and certainly Antarctica is only a very recent addition, to any degree of accuracy, to twentieth-century cartography. Hoax? Time travelers? Or merely precognition? Perhaps copies of maps drawn by visitors from outer space! But the fact that such questions are raised—even at the fringe of "acceptable" serious speculation about the nature of the world, space-time, and our place in it—is sufficient to put in perspective the speculations we are making now and will make in the rest of this book. The *really* crazy ideas don't come with the label "science fiction" at all; they come either in commonsense terms from fanatics who believe the most outrageous things, or from mathematical physicists probing the deepest mysteries of nature. Sometimes, the two sources overlap more than a little!

So far, we have two kinds of time travel paradox to puzzle over: the closed loop, which is more or less understandable in its own right (if not to "common sense"), and the new branch from the trunk of space-time. One way to "travel" into the future is also found commonly in SF—a form of suspended animation that enables the sleeper to survive a great span of centuries (a good recent example is James White's "Second Ending," in *Monsters & Medics*). But that doesn't come within our simple understanding of time travel—it's really no more than our everyday travel at a rate of 24 hours per day, but slept through, and with no way back.

Equally, time travel that simply reverses the "arrow of time"—so that the travel is all "backwards," but still at a steady 24 hours per day—doesn't seem worthy of the name time travel in the sense with which we're concerned. Philip Dick made a halfhearted effort to unravel the implications of such travel in his *Counter-Clock World*. But he managed at once to go both too far in his speculation and not far enough. In Dick's counter-clock world, time runs backwards, so that the dead are reborn and meals have to be regurgitated, not eaten, while cigarettes grow larger as they are smoked. The effect is supposed to be local and variable—different on Mars than on Earth, for example—which places it firmly in the fantasy category as far as our present understanding of space-time is concerned. (Modern physics doesn't suggest reverse time is necessarily impossible, but only envisages a reversal for the whole Universe—the whole of space-time.) But more to the point, the characters experiencing the reversal are aware of what is going on—they still think "forward" and plan "ahead." If time itself were reversed, the whole pattern of causality would be inverted, and the characters would never know anything odd was going on. In failing to allow for this, then, Dick doesn't go far enough; yet he goes too far in suggesting that anything noticeably odd would result from the reversal, in everyday terms. The only proper way to tell such a tale would be to write an ordinary story, then have it printed backwards. But maybe that would reduce its sales potential.

Of the two kinds of puzzle that seem fundamental to the concept of "real" time travel, the idea of a "splitting" of time and a development of parallel universes is the most intriguing, and the most relevant to any further developments. The closed loop idea, like the loops themselves, is a dead end—fun, but no more. Nevertheless, I don't want to leave the concept without some mention of three stories that really justify being called SF classics, and that just about squeeze the closed loop theme dry.

Keith Laumer's *Dinosaur Beach* cannot be described, but its wheels-within-wheels and loops-within-loops plot makes it essential reading for anyone puzzling over time travel paradoxes. More easily related, but no less puzzling, are the themes of two Heinlein tales, "All You Zombies" and "By His Bootstraps."

In the earlier, perhaps more pleasing variation on the theme ("By His Bootstraps") the hero is snatched into the future through a time "gate" operated by a mysterious old man. After some entertaining activities involving a doubling back through time so that the hero overlaps with himself (more than once), he eventually uses the time gate to set up what amounts, by the standards of America in 1941 (when the story first appeared), to something of a paradise on Earth—for male chauvinists—established at some unknown future date. Eventually, the circle is completed when the hero, now older, accidentally sweeps up in the time gate—you guessed again!—his own younger self, thus setting in motion the whole train of events.

The second story is less entertaining in itself, but even more complete as a time loop. In "All You Zombies" the narrator not only recruits himself to the time service but also—. No, on second thought I'll leave that one for you to investigate yourself; and we'll return to the theme or set of themes that form the basis of any serious investigation of time travel—the idea of parallel universes and the possibility of a sequential causality that follows its own logic, but not necessarily the logic of our everyday routine today.

Temporal Pigeonholes and the Cosmic Postman

The eminent astronomer-professor Sir Fred Hoyle has a habit of dressing up his science fiction stories with bits and pieces of genuine scientific speculation and philosophy; and his ideas about the nature of time are dealt with at some length in the novel *October the First Is Too Late*.[7] The story is one of time travel, but a peculiar kind of time travel in which different parts of the Earth seem to be reliving different eras from history. The details are of no importance here; but the discussion of time, put into the mouths of the two leading characters, most certainly is— especially since Hoyle states in a prefatory note to the volume that "the discussions of the significance of time and of the meaning of consciousness are intended to be quite serious."

This discussion hinges around the contention that our image of time as an ever-rolling stream is "a grotesque and absurd illusion." Although we feel subjectively that time progresses steadily from past to future, in physics, says Hoyle, all of time has equal validity and there is no way to single out "the present" as something of special importance. The image he chooses as an example is of the

[7] A more low-key mention of some of the same ideas occurs in the nonfiction book *Ten Faces of the Universe*, Chapter 2.

Earth going around the Sun. In the everyday world of three-dimensional space, we need to specify the time in order to know where the Earth is in its orbit; but in the physicists' four-dimensional world of space-time, the whole history of the Earth is laid out as a spiral in four dimensions moving around the Sun and through time. "There's absolutely no question of singling out a special point on the spiral and saying that particular point is the present position of the Earth." So we have the obvious puzzle of why we should *think* that the present marks a special point on this four-dimensional spiral—and that is where the temporal pigeonholes come in.

Everything, says Hoyle, exists; everything that was and everything that will be are in existence at all times. It is our consciousness that gives us a feeling of history and of time passing, but this is just an illusion. It's as if all of the physical information about the events of space-time were contained in a vast array of pigeonholes, each holding a set of notes describing the contents of other pigeonholes, with the holes numbered in some sequence to distinguish them from one another. Imagine some cosmic postman looking into the holes and reading the notes they contain. He will find that the notes in any one hole will always be accurate when they describe the contents of holes lower down the pecking order, but vague and contradictory when describing the contents of higher-numbered holes.

In Hoyle's analogy, any hole the cosmic postman looks into is the present, and the correct events it describes are those of the past; the vague, incorrect statements are predictions of future events. Now comes the tricky bit. Suppose part of the contents of some of the holes represents "your" consciousness. We imagine that "your" awareness is switched on when the postman looks into a hole. You remember past events, speculate about the future, and regard yourself as in the present. But the postman may suddenly turn his attention to another hole higher or lower in the sequence. This may spark into life "your" consciousness at times years later or earlier—but you're still able to "remember" only the events from earlier holes than the one now being investigated. The subjective impression is still one of time flowing, of the past being gone and the future yet to come; but every consciousness state can be triggered and retriggered indefinitely and in any order. We may, on this theory, relive our youth and age repeatedly, tangled up in any old order of events, and never know it! We may simply be single facets of one all-embracing cosmic consciousness:

> Along one wall of our office we have one complete set of pigeon holes, all in their tidy sequence. Along another wall we have another set of pigeon holes. Two completely different sets. But there is only one light. It dances about in both sets of pigeon holes. Wherever it happens to be, there is the phenomenon of consciousness. One set of pigeon holes is what you call *you,* the other is what I call *me.* It would be possible to experience both and never know it. It would be possible to follow the little patch of light wherever it went. There could be only one consciousness, al-

though there must certainly be more than one set of pigeon holes.[8]

According to these ideas, put forward seriously by one of the best known of modern astronomers (who once, aptly, held the post of Plumian Professor of Astronomy and Experimental Philosophy at the University of Cambridge), we may be experiencing time travel—and space travel—all of the "time" without ever knowing it! The problem now becomes one of finding out which pigeonhole we are in, and making the spot of light that triggers consciousness move as we wish, while at the same time leaving clues of some kind in the holes we jump through in order to mark the route.

Against that intriguing puzzle must be set the apparent implication that if everything that may be already is, and is fixed in four-dimensional space-time, then free will disappears and we have no way of changing the fixed pattern at all. But all is not lost. All "things" may be fixed, in the sense that the sets of holes are established. But which set of pigeonholes will the spot of light play on? Which individual holes? And in what order? Are some universes merely dead ends, never activated by consciousness? Might we be able to make the jump into those universes? The questions become more intriguing the more we look at the theory—and, equally, the more it looks as if we'll have to discard our old picture of time as an ever-rolling stream if we are to get on with the business of time travel.

[8] *October the First Is Too Late.*

With these exciting possibilities in mind, it seems appropriate to draw back a while and take breath. I've been bandying about such phrases as four-dimensional space-time, but without as yet explaining the concept properly. Before we can look more deeply into the mysteries of free will, the possibilities of parallel universes and alternative sets of pigeonholes, it seems appropriate to catch up on what the physicists and astronomers today regard as commonplace: the stretching and squeezing of space and time implicit in relativity theory, the bizarre possibility of a backwards universe running alongside our own, and then the physical basis of the parallel universes concept. Armed then with the crazy ideas of the physicists as well as those of philosophers and science fiction writers, we may really be able to get somewhere in pursuit of the Holy Grail of time travel.

PART TWO

Physical Timewarps

CHAPTER
4
Elastic Time—
The Relativistic
Squeeze

The mainstream of modern scientific thought—fully accepted by all except a tiny minority equivalent to the flat-Earthers, who still don't believe that the Earth is round—now accepts that both space and time can be stretched or squeezed, that the concept of two or more events occurring "simultaneously" doesn't work in the real Universe, and that even the order in which events separated in space actually take place depends on just how you look at them. All this is sober, solid scientific fact, derived from the same theories that have been proved to work in such applications as the esoteric "atom smasher" machines of high-energy physics, the practicality of nuclear power stations, and the horror of the hydrogen bomb. To most people, however, the problem on being confronted with such ideas for the first time is not stretching space or time, but stretching the mind to accommodate the new ideas!

So where do we begin? It's not my intention here to provide an instant guide to relativity theory; the aim of this book is to look at the peculiarities of the time part of what we are learning to think of as space-time. If you want a very good nitty-gritty guide to the theories of relativity at an undergraduate level, I can strongly recommend *Spacetime Physics,* by Edwin Taylor and John Wheeler; if you want the story of relativity with a minimum of mathematics in highly readable form, there is nothing to beat William Kaufmann's *The Cosmic Frontiers of General Relativity.* Here and now, though, let's start by looking at the situation in more philosophical terms. Why should we be surprised to learn of the elasticity of time (and space)? It all depends on the concepts we regard as common sense.

To someone born and raised on Earth, who doesn't travel very much and believes only the evidence of his own eyes, common sense gives a very clear world picture. Obviously, apart from a few hills and valleys, the Earth is flat; the stars, Sun, and Moon rotate about the flat Earth; things tend to stay where they are unless you push them around, and once you stop pushing they fall to the ground and lie still. But now consider the commonsense world view of someone born and raised in a large space station orbiting around the Earth. As we all can appreciate today from watching astronauts on TV, in that environment it is very clear that things don't tend to fall in any direction, but drift along in straight lines until they bump into something; push something and it moves, but when you stop pushing it keeps moving at the same velocity until something else pushes it.

It took Newton to appreciate that the commonsense world view for an inhabitant of the Earth might not apply generally and to explain the apparent movements of Sun, stars, and planets in terms of a different world view—the commonsense picture of our orbiting astronaut. And it is a similar consequence of parochialism that makes the implications of relativity theory seem odd to most people today.

Common sense—everyday experience—tells us that moving clocks behave no differently from stationary clocks, and that a yardstick is always a yard long however fast it moves and however you measure it. But just as our everyday experience doesn't include floating around in space, neither does it involve rushing about at a sizeable fraction of the speed of light, 30 thousand million centimeters a second. And just as it took Newton to jump beyond the commonsense picture appropriate to life on the surface of a planet, so it took Einstein to jump beyond the commonsense picture appropriate to a life spent moving very slowly compared with the speed of light.

Relativity theory is founded on two rocks. The first is that three-dimensional space and the flow of time cannot be separated, but are better considered as a four-dimensional whole, space-time. The second is that however you move and wherever you are in space-time, whenever you measure the speed of light it has the same invariant, constant value. Of these two concepts, the second is the one that causes all the trouble to ordinary common sense, with its corollary that nothing can ever travel faster than the speed of light. Instinctively, most of us respond by asking what happens if an astronaut traveling away from Earth at three-quarters the speed of light ($\frac{3}{4}c$) fires a smaller projectile ahead of him with speed $\frac{3}{4}c$. Surely now the projectile moves at $1\frac{1}{2}c$ relative to Earth?

But the answer is no! And the reason is that our simple law of addition of velocities, which works so well at the speeds encountered here on Earth, is no more accurate than the simple approximation that the Earth is flat, which works so well over the distances most of us travel every day. The correct law of addition of velocities is best seen physically as corresponding more to the addition law of angles than to the addition law of lengths. Because there are only 360 degrees in a circle, if you add an angle of 200° to an angle of 200° you *don't* get an angle of 400°— there is no such thing. The analogy is not exact, but it shows that even in the everyday world there are things which don't add up in the usual sense of the word. Einstein's theory predicts—and experimental tests have now confirmed the accuracy of the prediction—that the appropriate addition rule for velocities is actually given by the formula

$$V = \frac{v_1 + v_2}{1 + (v_1 v_2)/c^2}$$

where V is the "new" velocity resulting from adding two velocities v_1 and v_2, and c, as usual, is the velocity of light.[1]

The Physical Facts

For those without the mathematical background to appreciate trigonometrical subtleties, don't worry. Look at what the equation means in physical terms. First of all, look at the bottom part of the fraction on the right-hand side of the equals sign. The speed of light, c, is near enough 30 thousand million centimeters a second (3×10^{10} cm s^{-1}), and we are dividing the product of the two velocities we want to add up ($v_1 v_2$) by the *square* of this huge number, 9×10^{20} cm^2 s^{-2}. Unless v_1 and v_2 are also very big, the result is absolutely tiny and insignificant, so the equation boils down to the old familiar $V = v_1 + v_2$, since the complicated bit on the bottom comes

---

[1] Readers who are familiar with trigonometry and the relationships between angles in general will recognize the similarity of the equation to the rule for adding up the "hyperbolic tangents" of two angles:

$$\tanh (\theta_1 + \theta_2) = \frac{\tanh \theta_1 + \tanh \theta_2}{1 + \tanh \theta_1 \tanh \theta_2}$$

And those cognoscenti will also appreciate that however much you keep adding, the tanh $(\theta_1 + \theta_2)$ you end up with must always lie in the range from -1 to $+1$, corresponding to the limited range of angles obtained by going one way or the other around a complete circle. The thetas (θ) can be as big as you like, and in special relativity the name "velocity parameter" is sometimes used to describe the equivalent variable; but the hyperbolic tangents (tanh) have a restricted range just like velocities in the real world.

down to 1. So the correct rule of addition of velocities does give the same answer as common sense if you move about slowly compared with the speed of light.

Now look again at the intrepid astronaut shooting out his fast-moving projectile ahead of him. This time, with v_1 and v_2 both $\frac{3}{4}c$, we have to use the full equation, which becomes

$$V = \left(\frac{\frac{3}{4}c + \frac{3}{4}c}{1 + \frac{3}{4}c \times \frac{3}{4}c \div c^2} \right)$$

The c's on the bottom cancel out, while the ones on top add up to give

$$V = \left(\frac{1\frac{1}{2}c}{1 + \frac{9}{16}} \right) = \left(\frac{1\frac{1}{2}c}{\frac{25}{16}} \right) = 0.96c$$

However much you add up velocities, the total is always less than c, though it can get as close as you like without ever actually reaching it![2] And the only way to get a velocity *bigger* than c is to start out with one—a velocity of $2c$ added to one of $\frac{1}{2}c$, for example, gives a total of $1\frac{1}{4}c$, which is certainly curious, since it is less than the $2c$ we started with, but is surely cheating. Or is it? There is a hint here of a loophole in relativity theory, which we will look at later on. But for now let's get away from the numbers and look simply at the physical implications of relativity theory, especially as they affect our concepts of time—

[2] The only way to get a *total* of c is to *start out* with a total of c. Imagine two beams of light racing past one another, each with speed c measured by an observer on Earth. If one of the light beams could "measure" the speed of the other, it would be not $2c$ but $2c/(1 + 1 \times 1)$, with the c's cancelled out as before; in other words, $2c/2$ or simply c.

physical implications that, it cannot be stressed too often, have been proven in direct physical experiments as unambiguous as Newton's observation that an apple always falls downwards, never up.

In the world of relativity, the world that would be common sense if we hurtled about at speeds close to the speed of light, anyone who keeps moving at the same speed in a straight line—constant velocity—is entitled to think of himself as being "at rest" with everything else moving "relative" to him. Such an observer notices curious effects on objects that are moving past him; measuring rods shrink as their velocity increases, and the passage of time recorded on a moving clock runs slower and slower the closer the clock gets to the speed of light. And, to cap it all, the mass of a moving object also increases the faster the object moves.

These changes, following clear-cut rules that, like the law of addition of velocities, cancel down to our everyday physical equations of motion for velocities much less than c, give us a neat physical picture of why the speed of light can never be exceeded.[3] The decrease in length of a mov-

---

[3] For those who are interested, the rules are:

$$(\text{mass of object moving at velocity } v) = \frac{(\text{mass of same object when stationary})}{\sqrt{(1 - v^2/c^2)}}$$

$$(\text{length of object moving at velocity } v) = \frac{(\text{length of same object when}}{\text{stationary})} \times \sqrt{(1 - v^2/c^2)}$$

$$\begin{pmatrix}\text{Time between ticks} \\ \text{of moving clock}\end{pmatrix} = \frac{(\text{Time between ticks of same clock at rest})}{\sqrt{(1 - v^2/c^2)}}$$

The relativistic conversion factor $\sqrt{(1 - v^2/c^2)}$ becomes zero when $v = c$; anything multiplied by zero is zero, anything divided by zero is infinite.

ing object squeezes it down to zero "size" as v rushes up to the speed of light; at the same time, the mass of the moving object becomes infinite; and the passage of time, measured by any clock carried along at c, dwindles to a halt, with infinite time elapsed between successive ticks of the clock.

The Speed-of-Light Barrier

To make any object speed up, it is necessary to give it a push—apply a force; and the amount of push needed to produce a particular acceleration depends on the mass of the object. To make a body with infinite mass accelerate you would need infinite force, which is impossible. So no body moving slower than the speed of light can ever get past the "speed-of-light barrier." But we are interested in the effects on time of movement at high velocity—the so-called "time dilation" effects. And this is where Einstein's theory shattered once and for all the homely concept of "time like an ever-flowing stream." The theory still says that time does flow ever-forward, but gone is the idea of a steady flow. Now, the rate at which time passes depends on how you are moving through space—space and time are inextricably tangled. Move faster, and the stream runs slower for you; and this makes it possible, without any shadow of doubt, to use the relativistic time machine as a means of moving into the future. Only one-way time travel, to be sure—and as such just a variation on the steady travel through time at a rate of 24 hours a day that we all do now—but still a genuine prospect of skipping down the centuries without ageing as much as the Earth, Sun, and Solar System.

Before going on to look at the implications in human terms, let's look at one of the classic experiments that proves that the relativistic theory really is a good description of how the Universe "works." In particular, we should check on how the basic principle that any observer is entitled to assume that he is at rest works out in practice. For, of course, if you rush off into space at high velocity I can sit here on Earth and say that by my standards your clock is running slow. At the same time, you are quite entitled to say that your rocket is at rest while the whole Earth rushes away, and that by your standards *my* clock is running slow. And we would both be right!

Nature provides us with a variety of different "clocks"—the radioactive particles that each disintegrate, breaking up into a collection of different particles, in a regular way characteristic of the type of radioactive particle you start with. Measurements in the laboratory indicate how rapidly each species "decays" if left to its own devices. And with the advent of the great particle accelerators now common in high-energy physics, it has been possible to take such well-studied radioactive species and whirl them around at high speed. Measurements made during such experiments show that particles treated in this way do indeed live longer—by exactly the amount predicted by relativity theory. The moving clock really *does* run slow.

But how does the particle view events? Circular motion

is tricky to deal with, so let's take the simpler example of a particle shot down a long, straight accelerator. Suppose that the track is 100 meters (10^4 cm) long, and that the characteristic lifetime of the particle is one ten-millionth of a second (10^{-7} s), which is the kind of lifetime some of the particles used in these experiments really do have. Even traveling at the speed of light, with no time dilation, such a particle could travel only 30 meters [$(3 \times 10^{10}) \times 10^{-7}$ $= 3 \times 10^3$ cm $= 30$ m] before literally disappearing, breaking up into a set of different particles. But now apply the correct relativistic time dilation factor, and shoot another such particle down the tube at, say, $^{12}/_{13}$ c. Without time dilation, it would travel $^{12}/_{13} \times 30$ meters, that is, 27.7 m. As measured by observers at rest in the laboratory, however, the particle's lifetime is stretched by a factor of 2.6; so it travels 2.6 times as far down the tube before decaying— a total of 72 m. If Einstein's version of common sense is correct, the particles will be found more than twice as far down the tube as they would be if everyday common sense is correct; and that is indeed what happens.

The particle, however, would have none of this. According to his frame of reference he is stationary, while the tube, laboratory, and experimenters all rush past at $^{12}/_{13}$ c; and his lifetime is the usual 10^{-7} s. *But the tube is now shrunk by its high velocity,* and by exactly the same factor of 2.6. So the particle still gets 2.6 times further toward the end of the tube than if everyday common sense prevailed and there were no relativistic contraction of length!

This is a simple example, but if the calculations are done correctly the balance is always maintained. Observers in different frames of reference, moving at different velocities, have their own individual pictures of the Universe and their own ideas about whose clocks are running slow. But when it comes to anything that can be measured at one point in space-time—such as the exact spot in an accelerator tube where a radioactive particle expires—they will end up agreeing with each other, although not necessarily for the same reasons.

One-Way Time Travel

Those who are frightened by relativity theory, and the way it sweeps our ideas about absolute space and time out of the window, have tried to find genuine "paradoxes" caused by application of the rules according to Einstein. But they have never succeeded. The hoariest of these paradoxes concerns two hypothetical twin brothers, one of whom sets out on a space voyage at high velocity. The brother left on Earth will say that he is at rest and the spacefaring brother is moving, and therefore the latter's ageing process is slowed down. The brother who has gone a-voyaging sees the situation exactly reversed—so which twin "really" ages more slowly? There is no paradox as long as the brothers keep traveling at constant velocity relative to each other. Each is entitled to his own opinion, and there is no absolute standard against which their opinions can be tested. The only way to find out which twin is older is to bring the spacefarer back to Earth, turning his spaceship around and rushing him home at the same high velocity. But the very act of turning the spaceship around changes the situation. Now there is no longer the symmetry that allows each brother to watch the other receding while seeing himself as being at rest. We *know* which brother has been turned around, and that is the one who will be found to have aged less when the two meet again on Earth. Observers are only entitled to treat themselves as at rest provided they do not *change* their velocity (accelerate). Indeed, in the standard example I have given, we "knew" who was doing the traveling all along, since only one twin climbed into a spaceship and was accelerated to high velocity in the first place! The symmetry of the situation is destroyed, and with it any prospect of a paradox.

In order to get a better grasp on the reality of the situation, look at some numbers. Set aside the problem of just how quickly acceleration can be achieved in practice, and send one twin on a voyage covering 25 light years each way, at a velocity of 98 percent of *c*. According to clocks back on Earth, the voyage takes 51 years, and the stay-at-home brother ages by that amount; according to the traveler's clocks, the voyage takes only 10 years[4] and the traveler ages by that amount—on return, he is 31 years younger than his twin brother! Things get even more interesting if we consider longer voyages, which effectively hurtle the voyager dramatically into the future. We are back in the realms of science fiction—although it may not always be fiction—and need to turn to the imagination of

---

[4] The traveler covers "50 light years" in only 10 years because, of course, as far as he is concerned the journey has shrunk to just under 10 light years, exactly as in the case of the radioactive particle shot down a tube.

SF writers to conjure up an image of what such a voyage might be like. But before we do, it seems appropriate to at least mention that travel at high velocity is not the only way to stretch time.

The Gravitational Squeeze

So far, all this stretching and squeezing of time and space comes out of the simpler version of Einstein's theories, so-called "special relativity," which deals only with motion at constant velocity. The hairier developments, the General Theory of Relativity, also deal with accelerations—changes in velocity—and with gravity; and in particular the theory develops the idea that in many ways gravity and acceleration are equivalent. Anyone who has gone up in a high-speed elevator—let alone a fast airplane—will know the feeling of increased weight that acceleration induces. Gravity is the key force that holds the Universe together, and an understanding of gravity is essential to an understanding of the cosmos, which is why Einstein was so keen to develop his theories to include gravity. But the cosmological implications are not our main concern here.[5]

The aspect of General Relativity that concerns our search for a way to travel in time—a genuine timewarp—is the discovery that a gravitational field distorts both space and time in its vicinity, and in particular a clock in a

---

[5] If you *do* want to know more about gravity and cosmology, see *White Holes*.

strong gravitational field runs slow in much the same way as a rapidly moving clock. If the gravitational field is strong enough, time effectively stands still, as it would for a clock traveling at the speed of light; and this is one way of understanding the phenomenon of black holes which has received so much attention recently. At the boundary of a black hole, gravity is so intense that time stands still; and nothing ever emerges from a black hole because it would take an infinite time on clocks in the outside world for any-thing at the boundary—including light—to break free.

Inside a black hole, time as we know it (together with space as we know it) ceases to exist, with implications that will be investigated in the next chapter. But for now, just stick with that one point—that a very massive, compact object such as a black hole, with an intense gravitational field surrounding it, provides the intrepid space traveler with the means to jump into the future not just once but re-peatedly, and without necessarily having to travel far and fast between jumps. Just diving a spacecraft into the region of strong gravity and swinging out again on the other side, the astronaut would see time in the outside Uni-verse speeded up, with millennia—or longer intervals—flicking by in the few weeks he spent maneuvering his spacecraft around the hole and back again. A one-way ticket to eternity.

Worlds Out of Time

The idea of using the high-velocity time dilation effect to make this kind of voyage in time has been used by

many SF writers to good effect. Apart from anything
else—as the example of the traveling twin showed—this is
the only way to get to the stars and back within the span of
a human life (from the traveler's viewpoint, that is, since
many generations might pass on Earth during his travels).

Author Larry Niven uses the theme as well as anyone in
his "Tales of Known Space" series of stories. But he goes
further than his colleagues and predecessors in the genre
by making use of the gravitational time dilation effect as
well in his excellent novel *A World Out of Time*. Anyone
fascinated by the idea of relativistic timewarps should read
this book. The story involves both kinds of relativistic
timewarp, with the principal character involved traveling
at close to the speed of light on a voyage to the galactic
center, a journey so long that even with the aid of time
dilation Niven needs to invoke another SF standby—
a form of suspended animation called "cold sleep" enables the character to spend most of the journey in hiberna-
tion.

At the galactic center he finds—what else in a tale published in the mid-1970s?—a supermassive black hole, the
hub around which the whole of our Milky Way galaxy
wheels. This is where the plot thickens. Diving his com-
puter-controlled spaceship through the enormous gravitational field of the black hole, he swings the hurtling craft
around and back out toward the distant orbit of the Sun at
the fringes of the galaxy. But now the gravitational time
dilation effect has played its part, far stronger even than
the velocity time dilation of the imaginary journey. The return of the voyager takes him back to the solar system not

a few lifetimes after he left, not even a few millennia, but 3 million years in the future. And that is where the story really begins—the story of the Earth 3 million years from now, the world out of time. (So I haven't spoiled it for you by telling you about the time dilation effect—you should still read it!)

All of this is entirely possible. The kind of spaceship Niven describes is feasible (and I won't spoil your fun by detailing it here and now); there might very well be a black hole at the center of our galaxy; [6] and (admittedly stretching a point) a very lucky navigator might just be able to use the strong gravity field to jump into the future without getting his spaceship mangled by tidal forces in the process.

It is possible to imagine some very peculiar consequences of such travel becoming a regular thing, since travelers from any past age after its inception might pop up together or separately in any future time, in a bizarre mixture of backgrounds and philosophies. If you like the Niven tale, the place to look for further such complexities is in the "Forever War" series written by Joe Haldeman— the first story of which was, happily, published side by side with another Niven story in the January 1975 issue of *Analog*.

So at least some SF writers are now using imaginative extrapolation of proven scientific fact in time travel stories, instead of just saying, "He got into the time machine, set

---

[6] Or at least *some* kind of collapsed object; see *White Holes*.

84

the controls for 2179, and stepped out into the twenty-second century.'' The road to the future is now open, in principle at least. But what of the road to the past? Even within the framework of the conventional understanding of relativity theory, with time stretched and squeezed almost at will, the prospect of *reversing* the flow of time eludes us. Something new is needed to make two-way travel in time a possibility—something beyond the conventional understanding of relativity theory, but not necessarily beyond the broader implications of those equations that, although they have been studied now for decades, still have surprises in store, and still can be used as a prop for new and even more daring leaps of the imagination.

What we need is a loophole in the equations, a region where the conventional rules break down. And we have one—the black hole, inside which *anything* goes. ''Black holes'' has entered the common vocabulary of the Western world at least, and the layman coming to grips with this shattering new prospect has learned to think of them as literally holes in space. Yet Einstein's great triumph was in appreciating the unity of space and time; necessarily, a hole in space is really a hole in *space-time*. Or, looking at the other side of the coin from that of our commonsense physical intuition, a hole in space must also be *a hole in time*. Not only is there a loophole in the equations, but the equations themselves tell us where to find it. Within the standard framework of relativity, the time dilation effects discussed in this chapter constitute a real, but in themselves incomplete, form of time travel, which we might

85

call "time travel of the first kind." The next step, allowing the prospect of two-way journeys in time, is clearly something different—shall we say "time travel of the second kind"?

CHAPTER 5

A Hole in Time

The black hole as a hole in space has been the subject of considerable attention in books and the press since the publication of John Taylor's classic *Black Holes* in 1973. Where enough matter is gathered together in one place— and "enough" need only be the equivalent of a few suns like our own—it seems, according to the equations of relativity theory, that gravity becomes the ultimate "irresistible force," crushing the matter down into a literal point, a mathematical singularity, and crushing with it the very fabric of space-time. Before the singularity is reached, gravity around the superdense object is so strong that even light cannot escape (hence the term "black hole"); the region within which escape is impossible and ultimate collapse to the singularity is inevitable is bounded by the so-called "event horizon."

I have described the physical implications of the existence of holes in space, and the presence of space-time singularities in the Universe, in my book *White Holes;* the prospects that are, hypothetically at least, opened up by such holes for travel in space through a kind of cosmic subway have been elaborated by Adrian Berry in his own *The Iron Sun.* Here, of course, we want to know about the other aspect of this destruction of space-time by the intense gravity of a singularity, the prospects opened up for travel in time. But let's tread lightly, and attempt to broaden our minds a little for the astonishing concepts to come by looking first, briefly, at those implications for *space* travel—which are, in all honesty, boggling enough to the average mind.

Holes in Space

I don't intend to go over the background material that has been the subject of other complete books, since that would leave no space for the important new concepts here. Rather, I shall try to present the information in diagrammatic terms, using the principle that one picture is worth a thousand words, and taking advantage of the "Kruskal diagrams" used by relativists to describe the strange worlds of space-time singularities.

The problems of presenting diagrams on the flat, two-dimensional page to represent three-dimensional objects in space are of course nothing compared with the problems of presenting a two-dimensional image of events in four-dimensional space-time. But relativists get round this by squeezing all of "space" into one dimension, across the page, with the flow of time represented by the other direction, up the page. On such a diagram, an object that stands still in space "moves" only through time, straight up the page from bottom to top. An object that moved only in space and stayed always in the same time would "move" horizontally across the page—and such journeys are generally regarded as impossible. Choosing the standard definition of the speed of light in such a way that a light track would move at $45°$ across the page, the entire Universe is represented on the page in terms of possible and impossible journeys.

Such a standard Kruskal diagram, centered on the Here and Now, is shown in Figure 5.1. As we sit at the Here and Now, if nothing can travel faster than light, then the extent of everything that could ever possibly have influenced the Here and Now is contained in the region marked Past; similarly, if any effects we produce Here and Now spread out through the Universe at less than or equal to the speed of light, then we can only have any influence on the region marked Future. The rest of the Universe—fully half of the diagram—can have a real physical existence, but can neither influence nor be influenced by the Here and Now.

Looked at in terms of journeys, the situation draws a clear distinction between time travel and space travel. In this picture, a journey that corresponds to a line extending from the Here and Now into the Future, making an angle of less than $45°$ with the time line, is an allowable journey,

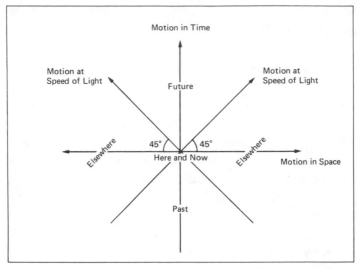

5.1   The standard space-time diagram used by relativists.
Credit: Geoffrey Wadsley

dubbed "timelike" because it is close to the time line. A journey making an angle of greater than 45° with the time line, however, is impossible according to the conventional understanding of relativity theory. Dubbed "spacelike," such a journey involves at least an element of what we would call travel in time. According to this understanding, we can never travel into the regions Elsewhere, and we can never travel back into the Past. Our possible journeys are limited not just to half, but to a *quarter* of the whole diagram!

But now let's bring a black hole into the picture. As the situation gets more complicated, the diagrams have to be

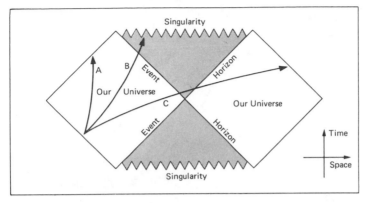

5.2 The space-time diagram of a simple, non-rotating black hole singularity. Credit: Geoffrey Wadsley

adapted to include the new complexities, and the result for the simplest kind of black hole is shown in Figure 5.2, a Penrose diagram of a nonrotating black hole. Here, the event horizon itself is defined by the lines corresponding to the speed of light (since, of course, this horizon is by definition the boundary corresponding to escape from the hole at the speed of light). The singularity in which space and time are destroyed appears at the beginning and end of time, and the entire outside Universe in which we live corresponds to "elsewhere" as far as anyone inside the black hole is concerned. This is just another way of saying that nothing can get out of a black hole of this kind.

Figure 5.2 shows three kinds of journey starting from a point in the outside Universe. Journey A is a conventional timelike journey that avoids the black hole and its singularity; journey B indicates the fate of a traveler who crosses the event horizon and plummets into the singularity, where he is crushed out of existence. Both of these are possible journeys by every standard test we can apply. But journey C is different. This one involves travel inside the event horizon and out again into another part of our Universe. It *looks* like space travel, but it involves speeds greater than that of light—it is a spacelike journey, forbidden by the rules of the Universe as we understand them.

All is not lost, however. For this simple picture applies only to the very simplest kind of black hole, one which sits quietly in space without moving—in particular, without rotating. That is a very unlikely state of affairs, because just about everything in the Universe rotates—planets, stars, and even galaxies. Whatever it was that collapsed down to form a black hole was almost certainly rotating, and as it collapsed the spin would go faster and faster, like a whirling ice skater who pulls in her arms toward her body. And a *rotating* black hole singularity is quite a different kettle of fish.

In physical terms, the rotation effectively distorts spacetime in such a way that it opens a gateway or gateways to other regions of space, to "elsewhere." In diagrammatic form we have the apparent confusion of Figure 5.3, with not one but two kinds of event horizon, the inner and the outer, and a checkerboard pattern that now seems to permit travel through the black hole and out again into a different part of our Universe, all the time along a legitimate timelike path. On this diagram, trips A, B, and C are as before; but the new possibility is indicated by journey D. Starting from one part of our Universe, the traveler passes through both outer and inner event horizons to reemerge in another part of our Universe, without ever exceeding the speed of light or being crushed by the singularity.

And this is where we begin to boggle the mind. For by "another part of our Universe" we don't just mean a different *place* but a different region of *space-time*. The emerging traveler is not just "elsewhere" but "elsewhen," past or future compared with his starting point, depending on the exact route he took around the singularity and through the black hole.

The Universe as Time Machine

This is where the specter of causality once again rears its ugly head. The "time travel is impossible" school draws back in horror at such a prospect and says that there "must be" something wrong with the theory that produces such an astonishing possibility. But such a position has no more foundation than the argument of flat-Earthers that the Earth "must be" flat or the Australians would fall off. Perhaps the theory is wrong, and it is impossible to take a shortcut across space-time using a convenient rotating singularity. But if so, it won't be because we say "I refuse to believe it," any more than the Earth is made flat by those who refuse to believe it is round. The paradoxes introduced by allowing a breakdown of causality are paradoxes in terms of the human mind, which does not experience them routinely and therefore cannot cope with them, just as many people today still cannot cope with the implications of elastic time and space discussed in Chapter Four, and now proved a reality.

In certain special circumstances, the theory tells us, causality as we know it might be violated. The theory told us before that in certain circumstances quite different from those of our everyday lives, time and space can be squeezed and stretched. That doesn't mean we need to take conscious account of elastic time from day to day; and the new implications may not involve taking constant account of noncausal behavior. But if we dabble with rotating singularities, watch out.

Two examples—one lighthearted, the other serious academic stuff—might help to flesh out some of these bones. The first was labeled science fiction, and appeared in the pages of the anthology *Faster Than Light,* edited by Jack Dann and George Zebrowski. Author George Martin describes a scientific expedition that travels through a hole in space-time to set up a research establishment in what seems to be an empty region of the Universe. The experiment, tinkering with ill-understood fundamental forces, is dangerous—hence its location. Naturally (or there wouldn't be much of a story) the experiment goes wrong; the artificial "star-ring" runs wild, and the scientists are withdrawn just before it rips space-time apart in a vast expanding explosion. What have the scientists created? A quasar, or something even more exotic, in a remote corner of the Universe? Or perhaps something even bigger—the Big Bang in which the Universe we know began—through

From the ground Stonehenge appears to be a
mysterious, awesome temple to an unknown god.
Credit: B. T. Batsford Ltd.

From the air Stonehenge's true nature as a clock of
the seasons becomes more apparent. The small outer
circle of round markers is more significant to
Stonehenge's use as an eclipse computer than the
striking but mathematically cruder central stones.
Credit: Aerofilms Ltd.

Solar eclipses, as studied by modern astronomers, are part of the natural running of our Solar System. To the ancients, however, they were terrifying acts of the gods—until a developing understanding of the rhythms of time associated with such events made it possible to <u>predict</u> them, thus seemingly bringing them under the control of the astronomer-priests. Credit: M. J. de Faubert Maunder

Symbol of Eastern philosophy, the Buddha. This fourteenth-century Tibetan sculpture is a beautiful work of art—but we also have much to learn from the East about the nature of time and its relation to the Universe. Credit: Crown Copyright. Victoria & Albert Museum

The pyramids of Central America, like those of Egypt, show clear astronomical alignments, evidence of the obsession of developing civilizations around the world with the cycles of the heavens and the nature of time. Credit: Mexican National Tourist Council

◄The archetype of modern man's obsession with accurate timekeeping—the Time Service Control from which the Greenwich time signals originate.
Credit: Royal Greenwich Observatory

This universal ring sundial provided a sophisticated means of keeping track of time anywhere in the Northern Hemisphere, when suitably adjusted. The Sun's rays pass through the adjustable pinhole to light a spot on the hour circle. Credit: Crown Copyright. Science Museum, London

Early civilizations lacked nothing in intellectual power and skill compared with today, although their equipment may have been different from the modern astronomer's. This column sundial is adjusted to the correct date, and then the time is shown by the shadow of the dragon's tail, with all the accuracy of a modern clock. Credit: Crown Copyright. Science Museum, London

Albert Einstein, born in 1879, was responsible for the twentieth-century revolution in understanding of the nature of time. Credit: Wide World Photos, Inc.

Elementary particles and the effects of high-speed travel on their local time frame are studied in facilities like this one at CERN in Geneva, used for experiments with pions and kaons. Credit: CERN

Although seemingly floating in space, the astronaut
is falling around the Earth in a closed orbit at thousands
of miles an hour. As he passes over all the time zones
of the world several times a day, his own "calendar" is
kept in step with that on the ground by the International
Date Line system (although in practice astronauts
keep the same "time" as mission control). But Einstein's
theory of relativity tells us that time passes slower,
by a tiny amount, for the moving astronaut—during the
mission he "ages" a few milliseconds less than his
colleagues on the ground. Credit: Wide World Photos, Inc.

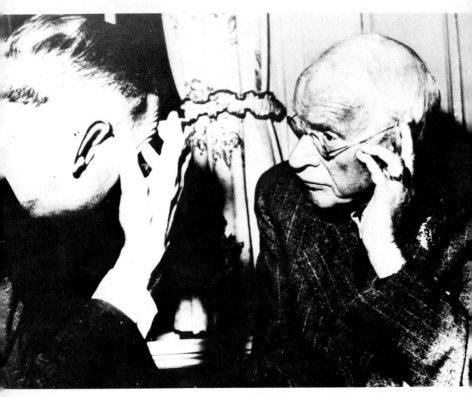

C. G. Jung (*right*), best known as one of the fathers of psychiatry, had a lifetime fascination with the nature of time and especially its relation to the Eastern philosophies contained in the *I Ching*.
Credit: Wide World Photos, Inc.

tampering not in the far reaches of space but in the far reaches of time? [1]

The ultimate paradox, if it could happen. But just because it is a paradox in everyday terms, does that make it impossible? One of the greatest mathematical geniuses of the twentieth century, Kurt Gödel, might not have thought so, judging from the evidence of his own work. Although best known for other studies, Gödel also looked at Einstein's field equations of General Relativity and in 1949 worked out "solutions" to those equations that include the effects of rotation. Each solution describes a possible Universe or set of Universes, one of which presumably corresponds to that in which we live. [2] A rotating Universe shares many features with a rotating black hole singularity, described by the same basic equations of relativity theory. Not least among these similarities is the discovery that the "Gödel universe" contains, in relativistic jargon, "closed timelike world lines." A look at Figure 5.1 will quickly show that a closed journey involves traveling back into the past somewhere along the line, as well as crossing the boundaries into elsewhere. In the Gödel universe it is possible, though not necessarily easy, to travel in time back into your own past.

[1] The story breaks down, of course, because "before" the Big Bang there would be no space and no time in which the scientists could come along to do their tinkering. As is so often the case with people used to thinking in everyday terms, George Martin equates "empty space" with "nothing," and doesn't realize that a real "nothing" has not even any space-time fabric in which anything, even empty space, could exist. But that doesn't stop it from being an entertaining story.
[2] See *White Holes*.

The reaction of the relativists to this discovery is that "obviously" Gödel's solution of Einstein's equations is not to be taken seriously—it must be some odd quirk of the mathematics, since "of course" time travel is impossible.[3] But the fact remains that the best description we have of the Universe is provided by General Relativity. Many people have tried to supplant it, but none have yet found a theory that works better than Einstein's wherever comparisons can be made. The very best theory of the Universe that we have tells us clearly and unambiguously that singularities associated with rotation allow for the possibility of time travel, whether the singularity is a local black hole or the Big Bang from which the Universe itself sprang. Until we have a better theory—one that is *proven* to be better by beating General Relativity at all the tests Einstein's theory has already passed—there is surely no reason to say it must be wrong, and every reason to look very carefully at its further implications, beyond those we are comfortably able to accept.

Holes in Time

Figure 5.3 tells only part of the story. This checkerboard pattern should really be imagined as extending infinitely far up the page and infinitely far down, weaving together an infinite number of regions of space-time. And, furthermore, these need not all correspond to our own Universe.

[3] See Paul Davies's *Gödel and General Relativity*.

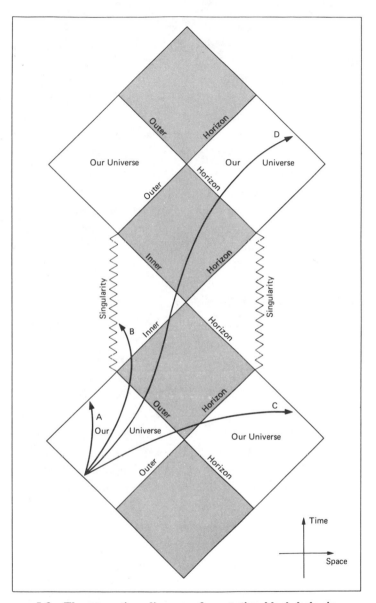

5.3 The space-time diagram of a rotating black hole singularity. The rotation opens up gates into other parts of our Universe, and other universes, which can be reached *without* exceeding the speed of light, but 'instantaneously,' as far as the traveler is concerned. Credit: Geoffrey Wadsley

Rather, the intrepid traveler along route D will find he has reached a different universe altogether; and by following a similar route ever further up the page he can visit as many different universes as he wishes—as long as his fuel holds out. This poses some interesting practical problems for anyone wanting to use a tame rotating singularity for travel in time *or* space. There may be no obvious way to get home! Repeated dives through the space-time gate may bring you into many mysterious places, some corresponding to the past or future of our own Universe, but many not. Just by taking pot luck, the chance of actually getting back to where you started must be infinitesimally small—and perhaps that is nature's way of avoiding the causality problem, since something that is possible in theory but infinitely hard to achieve in practice needn't worry the most ardent causality lover. The journeys into "elsewhere" and "elsewhen" could bring you unimaginable adventure, but never perhaps a homecoming.

Just how unimaginable the events of such a journey might be is brought home by the realization that a correct mathematical interpretation of the way space-time is distorted by a black hole tells us that inside the event horizon the roles of space and time are swapped over—the lines of "constant distance" point in the spacelike direction, and those of "constant time" in the timelike direction. Anything we associate with distance outside the horizon behaves like time inside—and if you can imagine what that means, you've got a better imagination than I have.

Another way to try to get a physical idea of what goes

on in these bizarre regions of space-time is to think of the rotating black hole as dragging the fabric of space around with it as it spins, like the water swirling around a plug hole. Straight lines through the space-time near such a hole become four-dimensional bent helices; and these are the paths followed by light rays, or astronauts, passing close to the singularity. But I'm not sure that this picture is really much help either. Let's go back to our space-time diagrams, and concentrate on the possibility of a journey that takes the intrepid astronaut back into his own past, as shown by Figure 5.4.

All the rules are obeyed—the journey never exceeds the speed of light, and the traveler never gets so close to the singularity that he is crushed out of existence on the way. This is time travel by anyone's definition, in the basic sense that we mean when we use the term in everyday speculation or in science fiction stories. And this is where the whole business is dismissed as impossible by so many—even William Kaufmann, in his excellent book *The Cosmic Frontiers of General Relativity,* writes

> Causality simply states that effects occur *after* their causes. If a light bulb in a room suddenly turns on, it is reasonable to assume that somebody flicked the switch a fraction of a second earlier. It is absurd to suppose that a light bulb could now turn on because somebody ten years in the future flicks the switch. The idea that effects could occur before their causes is denied by the rational human mind.

99

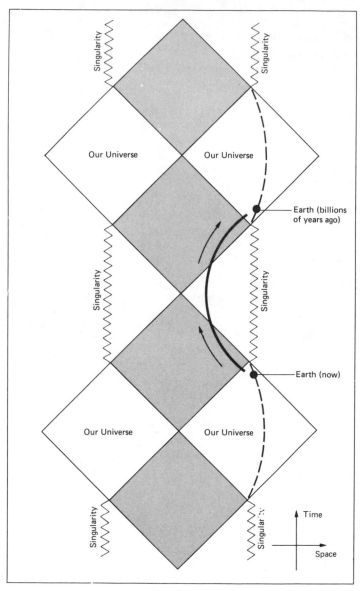

5.4 Using the black hole as a time machine. Travel near a singularity may bring the intrepid astronaut back to the same *place*, but at a different *time*. Credit: Geoffrey Wadsley

But is it? This is entirely a philosophical standpoint, one based on a belief in the nature of the Universe. It should be regarded in the same way as professions of a belief in God, or "rational" statements in support of agnostic views. The denial of acausal effects by the "rational human mind" tells us more about the human mind than about the physical nature of the Universe. And this brings us to an appropriate place to mention another topic that disturbs the orderly nature of the Universe as perceived by many cosmologists—again hinting that perhaps our neat human pictures are too simple.

White Holes

I have dealt at length with the implications of the existence of white holes—the opposite of black holes—in my earlier book *White Holes,* but to be thorough I should at least mention them now. A white hole, like a black hole, is associated with a space-time singularity, but one that matter pours out from in a cosmic gusher, instead of being swallowed up and squeezed out of our Universe. This brings in some obvious and pleasing prospects for symmetry—what falls into a black hole in one part of space-time can pop out somewhere, "somewhen" else and be spread around to form new stars and galaxies, which perhaps eventually collapse into new black holes and keep the cycle going. Indeed, within this picture the entire Universe as we know it may be the product of a white hole outburst, the explosive expansion we are used to thinking of as the Big Bang.

Although there is much more direct, observational evidence of exploding objects in our Universe (including the Universe itself!) than there is for collapse into singularities, many theorists have an innate dislike of the idea of white holes—perhaps because they are only just coming to terms with the implications of black holes, and don't want to move too fast too soon. There have been objections that such phenomena are mere "figments" of the equations (in which case, so is the whole Universe, since the same equations that describe the Big Bang describe white holes); and some calculations have shown that a nonrotating (Schwarzschild) white hole could get turned inside out by gravity, with a black hole forming *around* the white hole so that we could never detect it in the outside Universe.

But, as we found when looking at black hole singularities, the effects of rotation change the simple picture. In this case, the latest calculations, made by Dr. Kayll Lake of the University of Alberta in Canada, confirm that a white hole need not suffer this black hole fate. The equations say that white holes *can* exist in the real Universe; if you want to dismiss them as figments, you have to fall back on the same kind of claim that they are "unacceptable to the rational mind" used to dismiss what I have called time travel of the second kind.

Trading Space for Time—Profitable or Not?

All this, however, is rather removed from the prospect of practical time travel, at least in the foreseeable future. It may be that the very atoms in our bodies used to be in the middle of a large star in another part of the Universe (or another universe altogether), and that they fell through a black hole into a space-time tunnel, emerging from a gushing white hole at the center of our own galaxy to be removed into stars, planets, and even people. It is very likely that the best description of the origin of the Universe as we know it, the Big Bang, can best be understood in terms of some kind of space-time "travel" of this kind. But that still doesn't help us in trying to bend the rules a little in the region of space-time close to the here and now. To open a gateway to other regions of space-time you need, at the very least, a rotating black hole with a mass equivalent to several suns. Adrian Berry's rather ambitious solution to the problem[4] is to propose building a fleet of robot spaceships that will in turn "build" an artificial black hole by scooping up and gathering together the dust and gas from interstellar space. Even his superoptimism, however, places the grand project two or three centuries into the future—no more consolation here and now than the science fiction stories which invoke "hyperdrive" to provide instantaneous space travel.

By and large, though, those SF stories have the same failure of imagination that seems to have hit Berry in describing his artificial black hole as a transportation system—a "space machine," if you like. For, just as a breakdown of space-time in a black hole implies the possibility of time travel as well as space travel, any kind of

[4] See *The Iron Sun*.

103

"instantaneous" or faster-than-light (FTL) space travel must carry with it an element of time travel. And all too many of the sagas of time travel in SF also ignore the links between space and time—it is equally true that travel in time, if it can be achieved, must also involve travel in space! The whole business is a journey in space-time, and the problem is where to make the trade-off between space and time in order to get where you want to be at the right time.

Look at a simple example: a hypothetical time travel adventure in which our hero steps into his time capsule, travels a mere six months into the past, and steps out again. What will he step into? Clearly, empty space— since six months ago the Earth was on the other side of the Sun in its year-long orbit! A time machine must also be a space machine if our hero is to get where he wants.[5] But the problem does not end here. Another means of traveling in time might involve swapping one of the three dimensions of space for the fourth (time) dimension, the sort of thing that is achieved in a certain kind of SF adventure by a "simple rotation of the fourfold continuum," or something along those lines. Then, our hero might simply march down the time direction to his desired exit year and rotate back out into normal space-time. Or could he?

Space and time may be interchangeable, but if so the conversion factor used must surely be the speed of light—30 thousand million centimeters for every second. Rotate the continuum to make the time direction accessible

[5] Christopher Priest has used this concept to good effect in his weird parody of H. G. Wells, *The Space Machine*.

to space travel, and you'd surely have to travel 30 thousand million centimeters in order to get one second into the past or future—hardly an encouraging prospect. You still need FTL travel in order to move in time at any reasonable "rate"!

So FTL travel implies time travel, and time travel implies a kind of FTL space travel. James White developed the theme in his story "Tomorrow Is Too Far":

> It was not time-travel in the accepted Wellsian sense, where a few decades of travel into the future placed the operator in the same house and room that he had left, but in an older, perhaps ruined version of the same building, or where a similar jaunt into the past materialised them on rough ground before the place had been built. Time-travel stories of that kind had taken too much for granted.
>
> The present-day time-traveller had to be an astronaut as well. When he pressed the big yellow or red button he materialised in the past or future in exactly the same point in space—but in the meantime the earth, the solar system and the galaxy had either moved on or had not arrived there yet. . . . Time-travel involved no spatial displacement. It was just that everything in creation was moving in several different directions at once.

White's version of space-time travel does have its own drawbacks, and travelers who experience a "minus trip" (backwards in time) lose all their memories with a jolt that reverts them to mental babyhood. Of course, it's just a bit of fictional fun; but the point is a serious one, since *some* kind of psychological repercussions of such time travel are certainly in the cards. Especially if the whole concept of time as an ever-flowing stream is more a concept of the mind than of physical reality, as Hoyle and others have suggested (see Chapter Three, and Part Three later in the book), the new insights provided by changing the rules we live by are going to shatter more than one person's grasp on reality.

Using holes in space or distortions of the space-time continuum as a means of real, physical time travel—time travel of the second kind—remains in the realm of science fiction for now. Someday, though, we will certainly find a rotating black hole; and at the very least we will drop instrumented probes through the star gate, to emerge where they will. Will there be intelligent beings on the "other side," attracted by the singularity in their own region of space-time, who will pick up our probes and return their own, establishing two-way communication? Before we are able to make the jump ourselves, communication across space and time must be the most exciting prospect; and in fact we may not need to wait until we can tame a convenient black hole before communicating in this way. Indeed, by looking at one last subtlety of Einstein's equations we can see a loophole that might make such communication across the time barrier possible even now, and just might help to explain some of the puzzling mysteries of such phenomena as precognitive dreams.

Faster than Light—Backwards in Time

When we looked at the relativistic squeeze, the dilation of time produced by traveling at velocities close to the speed of light, we found several good reasons why the speed-of-light barrier can never be crossed. You can't start out with a speed *less* than that of light and go faster and faster until you pass the barrier, not least because time stands still at the speed of light itself, while length contracts to nothing. But it is only the actual speed of light itself that marks this forbidden territory. If, somehow, something existed that *always* traveled faster than light relative to us, then it too would behave in accordance with Einstein's equations, and it too would never be able to cross the speed-of-light barrier, being condemned to a permanent high-speed tour of the Universe. All this is allowed by those same equations that tell us how time and length can be stretched and squeezed by relativistic motion; for material particles, only the speed of light is impossible.

As ever, there are many people who argue that the possibility of such a world of high-speed particles beyond the light barrier is an "artifact of the equations"; that just because the equations allow for it doesn't mean it exists. But, as before, with a mathematical theory that has proved as successful as relativity, surely it makes more sense to believe in what the equations are telling us unless and until we find proof that they are telling lies. Some theorists, taking their courage in their hands, have done just this, looking at the implications of this FTL world, and giving the permanently FTL particles the name tachyons to distinguish them from the more mundane, slow-moving particles of everyday life, now dubbed tardons. Some experimenters have even tried to find tachyons by looking at the showers of particles that cascade permanently onto the Earth from space, the cosmic rays. And that is easier said than done, because these tachyons, if they do exist, live in a real Alice-in-Wonderland world, or perhaps a Looking-Glass world, where many "laws" of ordinary behavior are the opposite of what we know here and now.

The feature we are most interested in, of course, is the way time flows for tachyons. If we look at our tardon world, the faster a particle goes the slower time passes, until at the (impossible to achieve) speed of light, time is standing still. In the tachyon world, the same pattern applies, except that *time runs backwards compared with the flow in the tardon world*. It is just as difficult to approach the speed of light from the tardon side as from the tachyon side of the "barrier" —which in the tachyon world means that the easiest option is for particles to move faster and faster, hurtling backwards in time in the process.

When an energetic cosmic ray particle smashes into the top of the atmosphere, it produces a whole shower of "secondary" particles that can be detected on the ground by suitable equipment. Suppose some of these newly created particles are tachyons—perhaps they too can be detected on the ground. But, if they travel backwards in time, they will arrive at the ground *before* the original (primary) cosmic ray hits the top of the atmosphere—and even sooner before the conventional secondary shower is detected. This is what several experimenters have looked for—otherwise inexplicable blips on their detectors arriving shortly before ordinary cosmic ray showers. There is some evidence that these blips, produced by tachyon showers, do occur in some observations—not enough evidence yet to persuade the doubters, but certainly enough to encourage further speculations about what might happen if interactions between the tachyon and tardon worlds could be made to take place under controlled conditions.

The basic problem of communicating across interstellar distances would certainly be removed if we could send bursts of tachyons—instead of sluggish electromagnetic radiation, moving merely at the speed of light—from a "transmitter" to a "receiver." If tachyons really do exist—and the balance of evidence now is in their favor—it is only a matter of time before such a possibility becomes first a probability and then a reality. This may well offer a much more immediate prospect than any attempt to build our own black hole subway system, as well as offering other intriguing possible insights into the way the human mind interacts with the Universe around it.

The relationship between time and mind is the subject of Part Three of this book, and I don't want to get too far ahead of myself in trying to take a logical, ordered look at the nature of timewarps. But, just in passing for now, there does seem to be firm, clear evidence that at least on special occasions at least some people have experienced precognitive warnings of future events, often in the form of dreams. Is it possible that the human mind is able to interact with tachyons and that this could be a clue to the mystery of precognition? Certainly, stranger things have been suggested concerning the interaction between the mind and the world of elementary particles and quantum mechanics. Indeed, this hint of a quite different way of looking at the whole business of time brings us on toward new territory, beyond the confines of relativity theory applied in the Universe as we know it. The possibility that there might be an entire tachyon universe—where time runs backwards, waiting to be opened up to communication if not to actual travel—might sound as way-out as you can get and still have some tenuous links with the everyday practical world of physics. But this is nothing compared with the prospect of a whole array of alternative "parallel universes," offering prospects of travel neither forwards nor backwards but *sideways* in time. And this idea—time travel of the third kind—is not just way-out philosophical fantasizing, but in my view the most securely founded theory of all, offering the most real physical prospects for travel of a kind in time.

CHAPTER

6

Parallel Universes–
Sideways in Time

The possible existence of "alternative" versions of the Universe we know is a familiar theme from science fiction, and one we have touched on in Chapter Three. But today the concept of parallel universes and alternate worlds of probability is established not just in the musings of writers of fiction but solidly in the hard-core literature of mathematical physics. And, quite remarkably, one of the very first science fiction stories to develop the theme, four decades ago, provided a curious pre-echo of the scientific theorizing that was to come.

With such a pedigree, it seems appropriate to develop the theme with the aid of science fiction writers, whose works are in any case more intelligible than much of the mathematical physics that is now putting factual flesh on the bones of their fantasies. The author who gets credit for the first fully-worked-out alternate universe story seems to be Jack Williamson, whose *The Legion of Time* was first published as a magazine serial in 1938. The story itself is pretty awful—of the kind best described as "thud and blunder"—but the "what if" concept behind it was sufficiently outstanding to be remarked on at a lecture given by SF writer Harry Harrison in 1975:

> In *The Legion of Time* there appears . . . a concept different from the river-of-time theory. *What if* time is more like an ever-branching tree with countless possible futures? If each decision we make affects the future then there must be an infinite number of futures. In the river-of-time concept the future is immutable. If, on the way to work in the morning, we decide to take the bus instead of the tube and are killed in a bus accident, then that death was predestined. But if time is ever-branching then there are two futures—one in which we die in the accident and another where we live on, having taken the tube. . . .[1]

In fact, Williamson's original story wasn't quite like that. The alternative *possible* future worlds resulting from a different choice of actions remained only possibilities in his version, until a key choice of actions crystallized one possibility as "the" future. One of his characters explains the theory:

[1]*Explorations of the Marvellous,* ed. Peter Nicholls. Fontana, London, 1978.

112

> With the substitution of waves of probability for concrete particles, the world lines of objects are no longer the fixed and simple paths they once were. Geodesics have an infinite proliferation of possible branches, at the whim of subatomic indeterminism.[2]

It sounds like lovely gobbledygook to dress up a sagging SF story. But in reality it's a pretty fair instant description of the strange world of quantum physics, where the behavior of particles at the subatomic level is governed by rules of probability. The conceptual leap that Williamson just misses is the prospect that *all* of the infinite possibilities open to the future progress of the world might be equally real, producing an infinite array of universes every bit as real as "our" Universe, existing in some sense "parallel to" our own.

Let's go back to Fred Hoyle's version of the concept, the array of cosmic "pigeonholes" representing consciousness that we encountered in Chapter Three. A little later in his story, Hoyle has his theorist expound on what might happen if a Doomsday device were rigged so that purely at random the entire world is or is not destroyed:

> Do we all survive or don't we?
>
> My guess is that inevitably we appear to survive, because there is a division, the world divides into two, into two completely disparate stacks of pigeon holes. In one, a nucleus undergoes decay, explodes the bomb, and wipes us out. But the pigeon holes in

[2]*The Legion of Time.*

that case never contain anything further about life on the Earth. . . . In the other block, the Earth would be safe, our lives would continue. . . .[3]

In this tale, remember, Hoyle has specifically told us that while the story is fiction the ideas about the nature of time are intended to be taken quite seriously. The concept of an infinite variety of other universes is startling—but is it really any more startling than the idea of one infinite Universe, or than the concept of a unique "beginning" to that Universe in the Big Bang? Certainly the possible existence of other universes offers a neat resolution of the standard time travel paradoxes, since in an infinite array of alternative universes there must be an infinite variety of worlds differing from our own only in some tiny, subtle detail. Physicist Jack Sarfatti has summed the situation up:

We avoid the known paradoxes of time travel because of the many possible universes. A time traveler will probably return to a universe that is different from, but very similar to, the universe from which he started. These different universes usually differ in very subtle ways so that unless the time traveller is very observant he may not even realize he has returned to a different universe.[4]

[3] *October the First Is Too Late.*
[4] See Sarfatti's contribution to Bob Toben's astonishing book *Space-Time and Beyond*. If you think the ideas I am discussing here are weird, that book will show how relatively sober and unspectacular they really are! Strongly recommended for anyone interested in the philosophy of time travel and the interactions of the human mind with the Universe or universes.

114

Science, it seems, is only now catching up with science fiction in this area. But it isn't just science that has lagged behind; the world of literature has been equally sluggish. In 1976, SF fan and mainstream novelist Kingsley Amis, published a rather mundane parallel universe story called *The Alteration,* set in a world where the Church of Rome never lost its dominance over England and the rest of what we know as Protestant Europe. Mundane and obvious the story may have been to SF readers, who had been led a long way since Jack Williamson's breakthrough in the 1930s, but to mainstream reviewers commenting on the latest from author Amis, the concepts were seen as brilliant and original—resulting in reviews that must surely have embarrassed Amis himself, who was certainly aware of the much better treatment of a similar parallel world in Keith Roberts's *Pavane,* published eight years before and unnoticed by reviewers to whom SF was still beneath contempt in the late 1960s. If you want a straightforward but superb introduction to parallel universes, *Pavane* is the place to start—although even here there is a twist in the tale to make you think twice about what you have read. If, however, you want to jump into the deep end with the definitive tongue-in-cheek ultimate exploration of the possibilities of an infinite array of almost-but-not-quite-the-same alternate probability worlds, then the book to go for is David Gerrold's classic *The Man Who Folded Himself.* Here and now, though, the time has come to look at the basic physics that, according to our present understanding of the nature of the Universe, makes all these speculations something more than "mere" entertainment, and a valid

115

background to deeper philosophies. Let's have a look at the quantum background to the business of parallel universes, and at the prospect of moving sideways in time, making full use of the old dictum that one picture is worth a thousand words.

Worlds Beside Worlds

The basics of quantum theory rest upon the particle-wave "duality" of matter. At a fundamental level (for very small particles) matter is indistinguishable from wave packets, and behaves like a collection of electromagnetic waves—little bundles of wave energy. The archetypal example is provided by the case of ordinary light, where physicists today are used to thinking of light *either* as a continuous wave *or* as a collection of little energetic particles called photons. In Newton's time, though, the debate about the nature of light ran fast and furious, with the two opposed schools of thought each able to produce incontrovertible evidence that light *is* a wave or, alternatively, that light *is* made up of many particles. Both were right; today we accept both sets of proof and agree that light is both particle *and* wave. It's just that sometimes the wave nature shows up more prominently and sometimes the particle nature dominates.

Look at Figure 6.1. This is an example representing a beam of light, moving in the direction marked by the arrow, and acting as a wave pure and simple. When the waves run up against a screen with a couple of holes in it,

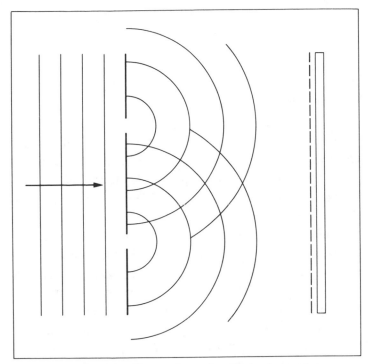

6.1 Waves moving through a pair of holes in an obstruction spread out like ripples on a pond to produce an interference pattern. This is one test which shows that light behaves, in some ways, like a wave. A light beam shone from left to right here, and passing through two pinholes in a screen, produces a pattern of light and dark shadows on a second screen. Credit: Geoffrey Wadsley

they spread out on the other side in concentric ripples like the waves set up when a pebble is tossed into a calm pond. The two sets of waves can now interact with one another to produce some strong peaks where they add together, and cancel out in other places to leave a shadow—the result is a pattern of light and dark as indicated schematically on the "screen" to the right of the figure.

Simple enough for ripples on a pond, or even light. But if the "holes" are small enough and the "screen" is sensitive enough, the same behavior can be observed for a beam of electrons—something we are definitely used to thinking of as made up of particles! In such an experiment, electrons are shot through a thin sheet or film of metal, where the gaps between atoms provide the "pinholes"; but the principle is exactly the same. The observed behavior is fine if you are dealing with waves, but how can it be explained in terms of particles? After all, for any one particle coming up to a screen with two holes the choice is simple—it goes through either one hole or the other, and carries on in a straight line, with no nonsense about "interference" with whatever went through the other hole.

Yet the experiment shows that with an appropriate arrangement of "screen" and "pinholes" a beam of electrons will produce an interference pattern exactly like that of Figure 6.1. Look at it now from the point of view of one electron, indicated in Figure 6.2 (or, indeed, consider the case of one photon in a beam of light!). Which hole does the electron "choose" to go through? The simple basis of quantum theory is, as Jack Williamson outlined so many decades ago, that a "choice" is made at random from all of the options open to a particle at the quantum level of electrons, photons, and weirder beasties. The choice may be weighted in certain directions, so that some "outcomes" are more likely than others in a given situa-

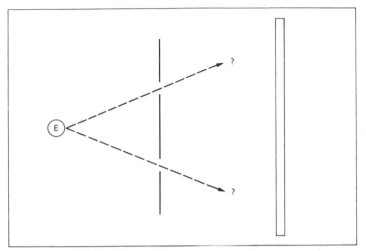

6.2 Electrons, too, behave like waves when put through the 'pinhole' test, though now the 'holes' and 'screen' must be on an atomic scale, the electron beam passing through a thin sheet of metal foil. But how does one electron 'know' whether it is to pass through one hole or the other? How can it 'interfere' with itself? It turns out that the electrons obey statistical rules, 'choosing' one option or another in this and more complex situations, at random, from among the alternatives available. Credit: Geoffrey Wadsley

tion. But *any* of the options might in principle occur. It's just that some are rather unlikely. The particles in my typewriter, or in this book, for example, *could* all happen to move in such a way that the typewriter (book) levitated off my desk. But it is vastly more likely that all the particles move in uncoordinated fashion as they jostle about, with a net result that the typewriter (book) as a whole doesn't move.

What does this mean at a human level? After all, our own actions are simply the sum of a great many "choices" being made at the quantum level every second. What quantum theory is telling us is that everything is possible, but most things are rather unlikely.

So, in human terms, replace the electron faced by two pinholes with a man faced by two doors. Which one will he choose to go through? Or consider the man faced by a fork in the road—will he go up or down? *Every* choice is possible, so why should any one actually be chosen to develop the progress of one unique "real" world into the future? The argument put forward so vividly in science fiction[5] and now being increasingly discussed by physicists and mathematicians is that whenever such a choice comes up the entire universe splits in two, as described so graphically by Fred Hoyle with his doomsday machine example, and *both* choices "really" take place, with two separate universes developing as a result. Look now at our hero faced with another choice, depicted in Figure 6.3, where the developing Universe is shown as a movie strip. Walking up to a chair, he may either walk past or sit down to read a book. What "happens" is that the Universe splits in two, producing a Y-shaped movie strip rather like a tuning fork—*both* courses of action are followed.

This process is now imagined to be going on all the time, at every level from the most subtle (the electron with a choice of two pinholes) up through the human (Shall the

---

[5] For a neat fictionalizing of some of the ideas coming up, see Keith Laumer's *The World Shuffler* and others in the same series.

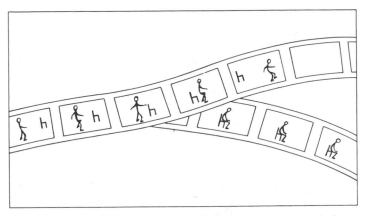

6.3 Some philosophers argue that every time a 'choice' of quantum alternatives is possible *both* options really occur, with the entire Universe splitting into parallel future worlds. Credit: Geoffrey Wadsley

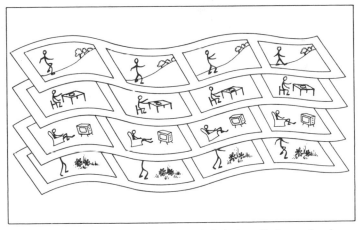

6.4. The parallel worlds of an infinitely split layered universe system in which everything that could possibly happen really has happened, some*when* across the sideways dimension of time. Credit: Geoffrey Wadsley

President press the button to start the ultimate nuclear war or not?) into the cosmic (Is a central explosion about to destroy our galaxy, turning it into a quasar?).

The net result is that our series of parallel universes might be imagined as an infinite *stack* of movie films (Figure 6.4). The layers nearest to "our" Universe (themselves infinitely deep!) differ only infinitesimally from our Universe, a boring succession of parallel universes so like our own that if we could visit them we might never know the difference. Leaving this local neighborhood and moving further sideways in time, we see in Figure 6.4 our hero living out four "lives" in four separate universes, the course of each life having been determined by a different pattern of quantum choices made some time in the past. Out walking in the hills; home eating; watching TV; gardening. *Anything* is possible and *everything* does happen in this world view, not some*where* but "some*when*."

This is all very well as an abstract concept for philosophers, or as the basis of fictional tales of entertainment. But the prospect of any real physical means of transportation—or even, more subtly, of a means of transferring information from one universe to another—opens up shattering possibilities. And however much we might like to stay in our cozy, self-contained Universe with no thought of what might lie next door, the nagging worry about where things go when they fall into black holes is alone enough to shake us out of this insular attitude. Jack Sarfatti again:

Singularities are entry and exit points of *that which is beyond space-time* projecting itself into space-time.[6]

So if you, or anything, can once travel through a singularity into the "outside," the choice of routes back into space-time—into the stack of parallel universes—is very unlikely to be restricted to holes going back into the Universe you started from. The implications of physical transportation sideways in time are shown schematically in Figure 6.5, where we are now looking at two very similar parallel universes, in which the same character is doing more or less the same things. If one of these "carbon copies" moves sideways in time, leaving an empty frame in our space-time movie, he could join his alter ego just a few blocks sideways in time. This is the idea at the heart of David Gerrold's *The Man Who Folded Himself:*

> If there was only *one* timestream, then paradoxes would be possible and time travel would have to be impossible. But every time you make a change in the timestream, no matter how slight, you are creating *another* timestream. (As far as you are concerned, it is the only timestream because you can't get back to the first one.) . . . You aren't really jumping through time, that's an illusion; what you are doing is leaving one timestream and jumping to—no, *creating*— another. The second one is identical to the one you have left, including all of the changes you have made

<hr>

[6] *Space-Time and Beyond.*

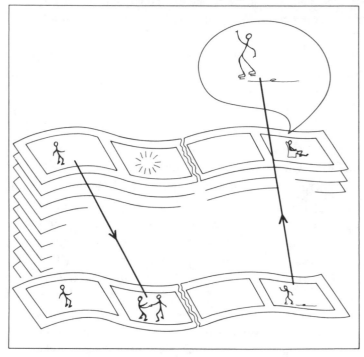

6.5 Can transposition across the parallel universe stack explain both paranormal dreams and even physical disappearances from 'our' world? Credit: Geoffrey Wadsley

in it—*up to the instant of your appearance.* At that moment you have changed the second timestream into a *different* timestream.

All time travel is now seen as implicitly involving travel sideways in time, even when the universe you end up in is so like the one you started from that you can't tell the difference. If this seems still too much to swallow as a physical possibility, try the scenario outlined on the right-hand side of Figure 6.5. Now, hero number 1 is dozing in his favorite armchair, and dreaming that he is out and about (judging from the figure, either skating or ballet dancing!). Where does the dream "come from"? Why, it's just a direct view, perhaps slightly distorted, of what alter ego, hero number 2, is doing in the nearby parallel universe!

This is an extreme, and rather silly, example. But then, where *do* dreams come from (especially those dreams that seem to foretell the future)? There are already great puzzles in the physical universe today, not least those connected with the elusive tachyons, the faster-than-light particles discussed in Chapter Five. It isn't much of a step from accepting the possible existence of particles that travel faster than light (backwards in time) to suggesting the existence of particles (waves) that can travel sideways in time, bringing to our subconscious vague impressions of another almost-but-not-quite-identical world. Déjà vu, anyone?

All this is bringing us very close to the realms of philosophy, and especially some of the ideas of Eastern philosophy and religions. Quantum mechanics and Eastern philosophy—a strange mixture. But not neccessarily so strange, as physicist Fritjof Capra has spelled out in his book *The Tao of Physics*.

Beyond Physics?

Capra's view is of the physical world of particles and the peculiarities of quantum effects in terms of particle behavior. He "sees" the rhythm in which particles are constantly being created and destroyed by natural processes as equivalent to "the Dance of Shiva, the Lord of Dancers worshiped by the Hindus." He talks of the "way"—or Tao—of physics in terms of the essential harmony of Eastern wisdom and Western science. But he stops short of the kind of discussion of the possible existence of parallel universes outlined here. Nevertheless, with Capra's insight opening a new window for us, we can also see the relation between some of these even more novel ideas of modern mathematical physics and the ancient ideas of Eastern philosophy. One quote will set the scene:

> The most important characteristic of the Eastern world view—one could almost say the essence of it—is the awareness of the unity and mutual interrelation of all things and events, the experience of all phenomena in the world as manifestations of a basic oneness. All things are seen as interdependent and inseparable parts of this cosmic whole; as different manifestations of the same ultimate reality.[7]

This seems to me an excellent way of viewing the existence of a multiplicity of parallel but related universes, all stemming from a different variety of "choices" at the quantum level—"different manifestations of the same ultimate reality"—so that no one time track can be viewed as the "real" universe, and all are equally valid.

Talking still in terms of *physical* effects—the space bit

[7] *The Tao of Physics.*

of space-time—Capra follows many recent physicists in remarking on the inseparability of an experimenter and his experiment in the world of quantum mechanics. The very act of observing a particle is now seen as influencing how that particle behaves—affecting which quantum choices it makes from the great variety open to it. Now, the observer is seen better as a "participator," whose presence influences that which is being observed:

> The idea of "participation instead of observation" has been formulated in modern physics only recently, but it is an idea which is well known to any student of mysticism. Mystical knowledge can never be obtained just by observation, but only by full participation with one's whole being. The notion of the participator is thus crucial to the Eastern world view, and the Eastern mystics have pushed this notion to the extreme, to a point where observer and observed, subject and object, are not only inseparable but also become indistinguishable.[8]

This is dangerous ground indeed. If we are all "participators" influencing the world about us at the quantum level by our very existence, and if the world is constantly splitting into a myriad of parallel worlds as different quantum choices arise, then in effect we are choosing which time track "we" follow by our reaction to, and influence on, events. David Gerrold's hero, creating his own new universes all the time, is nothing special after all—we *all*

[8] Ibid.

do this all the time, but we don't realize what is going on and, as far as everyday common sense is concerned at least, we cannot make the changes deliberately. We have here the beginnings of an insight into the mysteries of precognition, telekinetic ability, and so on. There is clear evidence that these bizarre events do sometimes seem to work for some people, and now we might see how.

Take the case of an individual with a mental ability to occasionally influence the roll of a pair of dice so that, say, the total seven comes up more often than by chance. On the participatory view of parallel universes, we say that every roll of the dice produces every possible combination of numbers, with the world splitting into parallel universes to accommodate them. Our "lucky" or "psychic" gambler has no influence over the dice, but is somehow directing his own consciousness along the world tracks where sevens keep coming up. The cosmic whole, the ultimate reality, is unchanged by his efforts, but his own perception of reality is focused on the "lucky" timelines. With everything that is possible really happening in some parallel time track, it is easy to extend this view to other "extrasensory" activities. The problem becomes not one of physical manipulation of "the" world, but of slipping sideways in time with the nonphysical self-awareness to find a world where things work out in the way we would wish.

Starting out from pure physics and moving through quantum mechanics, then, we are led to regard the human mind as of key importance in shaping the world as we know it—or rather the sum total of many human (and other) minds is the key factor; and the great majority of these minds are acting in a blind, uncoordinated fashion so that the world as we see it has the appearance of a largely random series of events at the quantum level.

Clearly, any real insight into the possibilities of time travel must involve a closer look at the relationship between time and mind, and this will be the subject of the remaining chapters of this book. But first, perhaps, we should tidy up our view of the modern physicists' understanding of the Universe (or universes) about us. True, as Capra reminds us, the Eastern mystics have a strong grasp of the inseparable nature of space and time, describing them as "interpenetrating," so that space cannot exist without time, nor time without space. And this is a pretty fair basis for coming to grips with the philosophical concepts of relativity theory. But we have now moved beyond this, to consider the possibility that at a fundamental level there is "really" no such thing as space. The Eastern mystics would tell you that in that case there is no such thing as time—and they would be right. So it is not that we have gone "beyond" physics after all, but that the new physics transcends the old, and transcends also, perhaps, the ideas of those Eastern mystics.

Transcendental Physics

Just how far beyond our everyday understanding of physics the experts have already ventured is shown by the work of cosmologists such as Paul Davies, who is among those who have discussed the speculation that while the Universe as we know it may have a definite beginning and a definite end, both "before" and "after" there have been and will be an infinite succession of other universes producing every possible combination of random fluctuations.

> Contained in this bizarre reasoning is the expectation that after a colossal number of different universes were produced in this way at random, the time would eventually come when the next fluctuation would produce an almost identical copy of the universe that we see now, complete with the Sun, Earth, Empire State Building and the reader! Of course, before a world indistinguishable from our own had been recreated, an almost limitless number of "near-misses" would first occur, some without the Empire State Building, even more without Africa, and many many more without the Earth at all.[9]

This is not a fashionable idea today, to say the least. But rearrange the pattern so that the infinite variety of universes run side by side instead of one after the other and you have a very different kettle of fish. Before the development of quantum theory the only way alternative universes could be envisaged was sequentially; now the possibility of an infinite variety of coexistent probability worlds must be as respectable as the concept of space-time—more so, according to one of the greatest of the pioneering modern mathematical physicists, John Wheeler. For, as

[9] *The Runaway Universe,* describing ideas originally put forward by Boltzman.

Wheeler sees it, the concept of "space-time" is as much an approximation in its way as the older Newtonian vision of the Universe.

> The uncertainty principle . . . deprives one of any way whatsoever to predict, or even to give meaning to, "the deterministic classical history of space evolving with time." That object which is central to all of classical general relativity, the four-dimensional space-time geometry, simply does not exist, except in a classical approximation.[10]

This doesn't mean that "classicial general relativity" is useless. It's a good, workable approximation as far as it goes, just as Newtonian mechanics is a good, workable approximation in everyday life. Probably (but we can't yet be certain) the working range of relativity theory extends to the ideas about black holes and space-time tunnels discussed in previous chapters here. But *certainly* at some deeper level it is correct to say "there is no such thing as space, nor any such thing as time." And that really opens the floodgates.

Paul Davies is optimistic, saying, "Doubtless a future theory will build together space-time and quantum theory in a much more basic way, and an entirely new concept of space-time will emerge."[11] But to what extent can these ideas of transcendental physics be regarded any longer as

---

[10] Quote from Wheeler reported by David Block in the *1976 Yearbook of Astronomy* edited by Patrick Moore.
[11] *Space and Time in the Modern Universe*.

physics rather than metaphysics? Biologist Lyall Watson, confronted by phenomena inexplicable in everyday scientific terms and indicating the existence of genuine "paranormal" powers, finds solace in the world of transcendental physics, where his physicist friends have taught him

> that the objective world in space and time does not exist and that we are forced to deal now not in facts, but in possibilities. Nobody in quantum mechanics talks about impossibilities any more. They have developed a kind of statistical mysticism, and physics becomes very hard to distinguish from metaphysics. And that makes things a little easier for a biologist faced with biological absurdities.[12]

To all intents and purposes the barriers between physics, metaphysics, and even science fiction disappear. Dr. Martin Clutton-Brock writes in a specialist astronomical article:

> The universe is supposed to contain everything, by its very definition: it is better, therefore, to talk about many *"worlds"*. We imagine the universe branching into infinitely many worlds, only one of which we experience. There are closed worlds and open worlds; initially uniform worlds and initially chaotic worlds; high entropy worlds and low entropy worlds. In most worlds, life never evolves; in some worlds, life

[12] *Gifts of Unknown Things.*

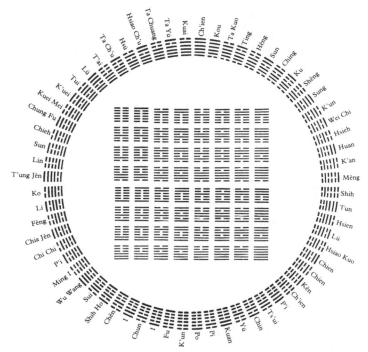

6.6 The sixty-four hexagrams displayed in the traditional arrangement of a circle and a square. Credit: Reprinted by permission of G.P. Putnam's Sons from *The Oracle of Change* by Alfred Douglas. Illustration copyright © 1971 by David Sheridan.

evolves but is scarce; and in relatively few worlds, life is abundant.[13]

While the scientist writes in these terms, the Eastern philosophers talk of a choice of paths and of finding the

[13] "Entropy per Baryon in a 'Many-Worlds' Cosmology," *Astrophysics & Space Science*, Vol. 47 p. 423 (1977).

"way"; and in his superb SF novel *The Man in the High Castle*, Philip Dick shows (surely not by chance) an alternative parallel present where the I Ching "oracle" plays a key part in everyday life, enabling the characters, if they wish, to choose, in effect, their way through the possible parallel worlds ahead. Only one character actually makes one brief journey sideways in time, and he does it not with the aid of any machine but in a trancelike state:

> Where am I? Out of my world, my space and time. The silver triangle disoriented me. I broke from my moorings and hence stand on nothing. Lesson to me forever. One seeks to contravene one's perceptions— why? So that one can wander utterly lost, without signposts or guide?

To that character, the prospect is frightening, a terrifying lesson never to be repeated. This echoes phenomena of the real world—precognition, dreams of future events, ghosts, and other "paranormal" phenomena. If something out of the ordinary leaks into our perception, we react by trying to sweep it away, ignoring it, or being frightened in some cases literally out of our minds. But conventional and unconventional physics have now taken us to the brink of an understanding of these phenomena and, perhaps, of ways to control them. Increasingly, it has become clear that the nature of time—and the nature of space—might be best understood not in terms of machinery or time-traveling "ships," but in terms of the most subtle structures of the fabric (if any) of the universes, and of the way in which they interact with human consciousness. When a physicist as respectable as Hoyle, schooled in the best scientific tradition, suggests that the way to understand the nature of time may be through looking at the nature of consciousness, at the way the flash of illumination flits about the myriad pigeonholes of the cosmic sorting office, and when the Wheelers of this world argue that there is no such thing as space-time, that in effect "it's all in the mind," then we have to accept the challenge in order to progress further in our investigation of timewarps. The physical possibilities discussed in this chapter seem to be complete—as far as they go. Beyond lies the still poorly understood realm of the mind, of the way we perceive the universes about us. And it is the links between time and mind that offer the possibility of, and an explanation for, genuine timewarping experiences that affect you and me.

PART THREE
Time and Mind

CHAPTER
7
Dreams and Reincarnation

The evidence that some form of ill-understood mental process is able to short-circuit the normal "causal" flow of time, and that this process works most effectively in providing some of us with "precognitive" dreams, is now compelling. It is not my intention here, in trying to assess the nature of the links between time and mind, to try to offer evidence that will persuade any doubters that something out of the ordinary run of everyday life does enable the dreamer to break the time barrier in some way. After all, this work has already been done for me, and there is nothing I need add to the evidence presented by, for example, Arthur Koestler[1] and C. G. Jung.[2] What we are

[1] See especially *The Roots of Coincidence*.
[2] See, for example, *Synchronicity*.

now able to do is to go to the next stage, accepting that to the mind operating without the rigid limits of "educated" patterns of thought time does *not* flow onward like a stream, and trying to relate this phenomenon to the realization of mathematical physicists that, after all, in the Universe at large the concept of time as an ever-flowing stream may be false. In a nutshell, I'm not concerned here with *whether* ESP, and especially precognitive dreams or visions, occurs—that case I take as proven. Rather, I want to know *how* (and perhaps *why*) such things occur.

But still, we need some idea of just what kind of phenomenon is going to concern us in the rest of this book. Jung's definition of what he terms "synchronicity" is a good place to start:

I defined synchronicity as a psychically conditioned relativity of space and time. Rhine's experiments show that in relation to the psyche space and time are, so to speak, "elastic" and can apparently be reduced almost to vanishing point, as though they were dependent on psychic conditions and did not exist in themselves but were only "postulated" by the conscious mind. In man's original view of the world, as we find it among primitives, space and time have a very precarious existence. They become "fixed" concepts only in the course of his mental development, thanks largely to the introduction of measurement.[3]

[3] Ibid.

These are the words of an authority on the psyche, and must be accorded the same respect as those of an authority on the nature of the physical world. Jung saw many "coincidences" as, in fact, meaningfully connected "acausal" events, crossing the conventionally understood barriers of cause and effect in time. One example he recounts is of a patient who was reporting to him a dream about a golden scarab when they heard a knocking at the window; when Jung turned and opened the window a beetle almost exactly answering the description "golden scarab" (it was actually a specimen of *Cetonia aurata,* the rose-chafer) flew in. Coincidence? Or a significant acausal event? (Acausal because the dream occurred *before* its cause, the entry of the beetle into the room.)

Telepathic Dreams

Psychologists, of course, have an obsessive interest in dreams and their interpretation, and the modern generation of psychologists provides us with evidence that must be taken as rather more compelling than that of pioneers such as Jung, if only because of the rather more strictly controlled conditions in which the evidence was gathered. Both straightforward telepathy and precognition seem the only explanations of some of this evidence, gathered most impressively in a book *Dream Telepathy* by Dr. Montague Ullman, Dr. Stanley Krippner, and Alan Vaughan. Unlike Jung, that other pioneer of psychology, Freud seemed reluctant to accept the evidence for telepathic dreams, probably because he never experienced one himself. But Dr. Ullman and his colleagues report a neat anecdote that indicates how Freud was persuaded, almost "against his better judgement," that such things do happen.

The convincing evidence came from a patient whose telepathic dream showed just the same Freudian distortions of everyday life as an ordinary dream. The patient, a middle-aged man, dreamed that his wife had given birth to twins; soon afterwards, he heard from his son-in-law that, on the night of the dream his *daughter* had given birth to twins, a month early and with only one child being expected. The substitution of wife for daughter in the dream fitted in so well with Freudian psychology as to convince Freud himself of the reality of the experience!

Telepathic dreams or waking telepathy are, of course, as relevant to our inquiry into the nature of time as precognition. Telepathy involves, in some sense, a cancelling out of space, or travel of a message faster than the speed of light, both different facets of the space-time puzzle. The Ullman/Krippner/Vaughan study uses both reports of telepathic dreams discovered from treatment of patients—including, intriguingly, several where the *patient* dreamed about a problem that had been worrying the *doctor*—and, most convincing of all, direct experiments in which volunteers sleeping under observation were able to dream about events related to pictures that were viewed by "agents" awake nearby, or even some distance away. In one example among many that they discuss, the target picture (not revealed to the experimental subject before he went to sleep) was Chagall's "Paris through a Window;" the subject dreamed of walking in the French Quarter, of a strange "Moroccan" or Spanish village (aptly fitting the picture, which looks more like such a village with the Eiffel Tower stuck in it than like the real Paris), and other related scenes.

This kind of detailed dream study was only made possible in the 1950s, when the links between periods of "rapid eye movement" (REM) and dreams in sleepers were first noticed. Most of us remember our dreams rather poorly, and dream several times during one night, so that it is a very confusing task even for a psychologist to unravel the nature of a previous night's dreaming even from a cooperative subject. But knowing that REM activity and associated changes in the pattern of the brain's activity occur during dreams, experimenters can now wire up a subject so that the onset of dreaming is immediately recorded. At the end of a dream period, the subject is awakened and describes the dream immediately, onto a tape recorder, while still in that cozy half-asleep stage in which such things are remembered best.

So it is that there are really only 20 years or so of reliable experiments with dream telepathy to draw on, and it is hardly surprising that it has taken until the 1970s for the evidence to become firmly established. This is true, at least, for telepathic dreams, since otherwise there is always the nagging doubt that something the subject has seen or heard between waking and reporting the dream may have colored the description. But the same doubts cannot apply to precognitive dreams, where the description is reported—even written down or printed—before the events that were dreamed about take place.

Precognitive Dreams

Dreaming about the future seems to stir up an even more emotional response among those who refuse to accept the evidence than telepathic dreams. Yet, as we have seen, anything that implies a message traveling "instantaneously," as in a telepathic dream, must also imply the possibility of travel backwards in time. And if we guess, for the moment, that telepathic communication involves some form of particle like the tachyon interacting with the human brain, what could be more natural than that the pattern produced by one brain should interact most strongly with the very same brain, but at a different time, so that future experiences are in some way communicated back into the past of the person who experiences them?

This possibility offers hope of explaining why precognitive dreams seem to be about dramatic events, such as the sinking of the Titanic, rather than about the minutiae of everyday life. Only a big emotional impact, perhaps, can produce a strong enough interaction to provide the wave of information traveling back through time on a train of tachyons. But still, some particularly sensitive people seem able to produce precognitive dreams, if not quite to order then very nearly so. A study of one such person is reported in *Dream Telepathy*—Malcolm Bessent, who (among other notable achievements) in the course of *one night* in late November 1969 forecast the following events that actually took place: a shipping disaster involving a Greek tanker owned by Onassis within four to six months; the death of General de Gaulle "within one year"; and a change in the government of Britain in the summer of 1970.[4] Under controlled conditions, Bessent was asked to try to dream about events of the next day, events made up of "target" pictures that were chosen at random from a pool *after* the dream reports had been transcribed and *before* the experimenter making the choice had access to the reports.[5] The results were clear-cut and impressive:

> Taken together, the two Bessent studies have lent experimental corroboration to the precognitive hypothesis. At least one psychic sensitive (Bessent) was able to dream precognitively fourteen out of sixteen times.[6]

Even in the face of such evidence, however, we must take some account of the fact that such experiments are designed by people in the context of the everyday flow of time. What the results tell us is that this everyday conception is wrong in some fundamental sense; but we can't learn any more than that until we have some alternative view of the nature of space-time to put in its place and test

---

[4]*Dream Telepathy*. These predictions were written down, witnessed, and registered with the Central Premonitions Registry in New York.
[5]Details of the experiments and precautions taken are given in Chapter 14 of *Dream Telepathy*.
[6]*Dream Telepathy*.

by experiment or observation. Although many people have become deeply interested in one aspect of the problem of timewarps, this narrow interest often leads them along narrow paths with little thought to the broader implications of their investigations, so that coming along afterwards the reaction is all too often "If only they'd done so and so" or "If only they'd asked this question." Nowhere is this difficulty more obvious than in reported studies of investigations of reincarnations—people who "remember" previous lives, either in the conventional sense or under hyponosis.

Past Lives

A standard stage trick of the entertainer/hypnotist is to cause a volunteer in a hypnotic trance to "regress" to childhood, when he or she behaves and speaks in a childish manner—the entertainment supposedly lying in the bizarre sight of an adult acting like a child. Some hypnotists, however, have gone further, and not in the guise of entertainment. They have found that it is possible to ask a trance subject to regress to a time before he was born; in some cases, this results in the subject being able to describe a past life or lives, more often than not assuming the voice, accent, and mannerisms of a completely different personality, perhaps even a different sex.

Clearly this is telling us something very deep about the human mind. But what? Are these genuine examples of remembering past lives, implying the reality of reincarnation? Or, as critics of this theory usually suggest, do the trance subjects act out fantasies, perhaps based on something they have seen or read but that is forgotten by the conscious mind? Either way the ability would be just as remarkable. But I now wish to raise a third, neglected possibility. The trance state bears some similarities to sleep, and we now know that the dreaming mind seems less fixed in time than the conscious, waking mind. Is it possible that under trance, rather than reliving past lives, these subjects are somehow able to scan far across the time barrier to *view* the past lives of *other people*? (Of course, we shouldn't forget that according to some philosophies, like that of Hoyle discussed elsewhere in this book, we may all be manifestations of one consciousness anyway!) With such a possibility in mind, we can relate such experiences more easily to those of dreams. Again, we see no objection to the idea that some minds may "tune" more easily to a particular trance subject than others, so that the close identification of these regressions occurs only with a few scenes displaced "out of time."

Such an idea is certainly an attractive explanation even of features of ordinary regressions in which subjects "remember" fantastic details about events from their babyhood, including in some cases their own birth. Does it make more sense to argue that such details really are stored in the memory forever, albeit inaccessible to the waking mind, than to argue that under trance the adult mind, with its full faculties, is able to tune in to such past events—especially something as traumatic as birth—and interpret what is going on with the basis of adult experience as a guide? If there is any hint at all that time can be warped, then to me the second alternative is much more plausible. And evidence there is, in plenty.

Let's look at a very few of the classic examples of regression in the light of this new hypothesis. Probably the single most widely publicized case is that of "Bridey Murphy." Virginia Tighe was an ordinary American wife and mother, born in Madison, Wisconsin, but living in Pueblo with her husband and three children when she met an amateur hypnotist, Morey Bernstein, and became interested in hypnotism herself. Finding herself a good, sensitive subject, she allowed Bernstein to place her in a trance state and was "regressed" to a time before her birth, the early nineteenth century. In this state, she described her life as Bridey Kathleen Murphy, born in Cork in 1798. She gave great detail about her parents, school, husband, and life, which ended in 1864.[7] A book published about this regression caused a sensation and fierce controversy. But all of the description of life in early nineteenth-century Ireland held up under scrutiny, and it is particularly striking that the slang or dialect expressions used by "Bridey Murphy" were typical of those common in the region up to about the 1880s, but far from common in the twentieth century and hardly something that an American would have picked up by casual reading over half a century later.

Suppose now, however, that instead of Virginia Tighe

[7] This brief synopsis is drawn from a more detailed account in *Journey to Infinity* by J. von Buttlar, one of the many books in which a fuller account of the Bridey Murphy case can be found.

being a "reincarnation" of Bridey Murphy, for some reason their minds were particularly attuned, that they
operated "on the same wavelength." Now, in the trance
state, we can envisage communication between these
minds being established just as a radio is "tuned in" to a
particular broadcasting station, so that the receptive, essentially unconscious mind of the trance subject Virginia
Tighe relays the images broadcast by Bridey Murphy in
the previous century, but which travel across the barriers
of time. Is this any less likely than the idea that the essence of a person is somehow reborn, with all the memories of past lives present but locked away in the unconscious?

Leaving aside the Bridey Murphy case, however, the
most impressive single-volume collection of regression evidence that I know of is provided by the book *More Lives
Than One?* by Jeffrey Iverson. This stems from a TV pro-
gram that Iverson made for the BBC about hypnotist Ar-
nall Bloxham, a specialist in "regressions," who had had
the foresight to keep tape recordings of the descriptions
recounted by many of his subjects. Among other things,
these tapes offer a telling counter to those critics who
argue that trance subjects are unconsciously providing em-
broidered tales of some dramatic past life in order to gain
attention. In Bloxham's own words:

> Most of my tapes are of deadly dull, ordinary people
> who have lived and died having done nothing whatso-
> ever—perhaps been housewives and nothing has ever

happened. The tapes people hear about are only the highlights of my researches.[8]

Is this the behavior of subjects seeking to gain attention—or indeed of people dramatizing on the basis of something read long ago? From the viewpoint of "telepathy across time," however, it is easy to explain—those people in the past may have led deadly dull lives, but their minds happened to be in tune with those of Bloxham's subjects—who, with respect, are in many cases themselves dull, ordinary people. A high percentage of the Bloxham tapes, however, relate to lives that ended in sudden death, however mundane they may have been until then. And is this so unreasonable? What is more likely than a sudden death to produce a mind trauma that sends its shock waves reverberating, in some as yet ill-understood fashion, along the corridors of time?

The tale told by Iverson (on behalf of Bloxham—now, alas, dead) makes gripping reading and is striking again for the attention to detail and accuracy of slang or dialect expressions, in regressions about the lives of people ranging from a Roman wife to a servant in medieval France and a gunner in the eighteenth-century British navy. But my chief interest in the book is where it goes beyond the usual intriguing narrative account of past lives. For, in the interests of his TV audience, Iverson was involved in further research and attempts to "reconstruct" regressions that were recent enough in time and local enough in space

[8]*More Lives Than One?*

(within the United Kingdom) for conventional historical/archaeological methods to be applied. The case chosen for particular study was first reenacted as a hypnotized regression in front of the TV cameras. Very briefly, the story told was of a "Rebecca the Jewess" living in York in 1189 and dying violently, having hidden in a crypt or cellar of a church in York, during a massacre of Jews in 1190.

The massacre certainly occurred, as of course the subject might have known from reading about it. But her description of events leading up to it, and of the final moments in the church, were so specific that the TV team was encouraged to try to locate the very church where it happened. This they did—except, unfortunately, that there was no trace of a crypt. Everything else tallied: the district, the name of the street, and so on. But St. Mary's, Castlegate, York, lacked this single additional essential, the underground room described by Rebecca. Then, six months after the investigation, Iverson heard from his contact in York:

> In September, during the renovation of the church, a workman certainly found something that seems to have been a crypt—very rare in York except for the Minster—under the chancel of that church . . . he had seen round stone arches and vaults. Not much to go on, but . . . this would point to a Norman or Romanesque period of building, i.e., before 1190 rather than after it. So, if one wanted to carry this argument to a conclusion, this crypt could be the place where, if you believe her story, Rebecca met

her doom. More certainly still, the discovery of reused Roman and Anglo-Saxon columns and masonry this summer below the present floor level in St. Mary's, makes it absolutely clear there was a church on the site in Rebecca's time.[9]

No way could that information have come out of any book accessible to the trance subject before her "regression"! But, again, the whole episode can be well explained by the idea of mental *communication* across time rather than that of reliving a past life.

Such a possibility also provides a foundation for the concept of "sensitivity" used by mediums, hypnotists, and even spiritualists—why shouldn't some people be better "receivers" of such images than others? And, a point taken up by Iverson, it may tackle the puzzle of why such regressions do not cause the subject to revert entirely to the language of a past life—Bloxham's "Roman wife," for example, did not speak Latin! There are recorded cases of regressions that did involve speaking long dead languages or dialects, but this is far from being true in all cases. Iverson points out that memories are often divorced from speech, and that if you describe now events that happened to you as a five-year-old you do not use the speech of a five-year-old. But to me this is only part of the story. The lives described during regression are *not* the subjects' "own" past lives but something they *observe* through a timewarping mental communication. In that case, while

[9]*More Lives Than One?*

slang or dialect approximating their own language might be presented "raw," as it were, something completely divorced from everyday experience could only appear in translated form. We do not, after all, "record" our impressions of life verbally, and there is no reason to think that they would be transmitted—by tachyons or whatever—in verbal terms, but rather as a full range of sensory images ready to be absorbed and interpreted by the receiver. The difference, if you like, between "analog" communication and "digital" transfer of information.

But one big frustration remains—the "if only" that I mentioned above. Blinkered by the established concept of reincarnation and the unspoken view that the future is yet to come and so has no form, Bloxham (like essentially all his colleagues involved in regression studies) has made no effort to investigate *future* lives. Yet, if there is anything in my hypothesis at all, the future should be as accessible as the past to the unconscious mind in the trance state. Will some hypnotist now take up the challenge and investigate—what? Pre-incarnation? Progression? The mental barrier has been largely, of course, the feeling that "free will" must imply that the future is yet unformed, and that we could only "know" about the future if all is preordained—a concept so unwelcome that most of us instinctively shy away from it, and fail even to investigate the possibility.

Now, however, we have seen that free will and a fixed future pattern of events may be quite compatible. If an infinite variety of possible future worlds exists, across a fan sideways in time, and we have the option of choosing by our actions which path across this fixed future landscape we will take, then precognition becomes much more understandable, and we begin to grasp the significance not only of those dreams and visions that "come true" but also of those that *very nearly* come true—we may be receiving information from one future world, and by using that information, perhaps, diverting our conscious flow into a nearby, but slightly different, parallel world. I hesitate to point out that the same can also be said of past worlds, so that any minor discrepancies between regressions and historical fact could be accounted for as the result of "scanning" not just backwards but also sideways in time. It's far too easy a cop-out for the slapdash investigator! Nevertheless, if I don't mention it someone else will—and this does not, of course, detract from such very close correspondences between history and regression as those of Rebecca. The future, though, is a much more exciting prospect—that, after all, is where we are going, unless we find a way to control these timewarps and bend the Universe to our bidding.

Future Worlds

Even within the more "conventional" view of a single thread of time, along which our consciousness moves in some fashion, there is no reason to suppose that echoes of future events may not reach back in time. Dr. Robert Thouless, in his foreword to Andrew MacKenzie's *Riddle of the Future,* reports a suggestion made by Dr. Russell Targ, of the Stanford Research Institute in California:

> He invites us to suppose that the single direction of time with information always flowing in the direction from the past to the future may not be a logical necessity but a firmly based habit of thought. Perhaps an event producing strong effects after its occurrence may also produce a weak effect earlier in time. A paranormal precognition might be such a weak effect, earlier in time than the event producing it. If we have the experience of waking up just before the alarm clock goes off, this might be an example of an event with strong aftereffects producing a weak effect earlier in time.

The strong aftereffects, of course, would not be the ringing of the clock but rather the mental shock produced in the hearer by the sudden noise—the communication through time would be telepathy from the listener's mind back to his own earlier, sleeping self. Even so, I remain wary of any hint that the future is predestined, and would prefer to elaborate this picture by suggesting that the shock of the alarm clock *waking* the hearer in one time track produces ripples both backwards and *sideways* in time to wake a *counterpart* a little early in a nearby time track.

The possibility of parallel universes also helps to resolve the puzzle of what appear to be "closed loops" in time. Another example reported by Andrew MacKenzie[10] con-

---

[10] *Riddle of the Future,* Chapter I.

cerns a woman who had a "vision" of a particularly striking patterned stone slab, and made a sketch of what she saw in the hallucination. She showed the sketch to a friend, who knew the stone in question and took her to a museum where she saw it, in the conventional sense, for the first time. Where does such a loop of cause and effect begin? In the picture of a unique time thread any attempt to explain such a phenomenon leaves the thread hopelessly tangled and the mind not a little confused—but suppose now that in some nearby universe, displaced sideways in time relative to our own, the woman had already visited that museum and seen the stone slab, making such an impression on her that she "broadcast" a picture of it across to her "sister" in our Universe. No paradox here, and no necessity of tangling the thread of time.

Nevertheless, the possibility of confusing the mind remains, even in the realm of parallel or "layered" universes. Professor Jack Sarfatti, attempting to shed some light on the various phenomena of ESP, comes up with the following:

> I believe the gravitational distortion of space and time predicted in Einstein's general theory of relativity provides a possible scientific explanation of precognition, retrocognition, clairvoyance, and astral projection, provided we accept the additional postulates that individual consciousness can alter the biogravitational field of a living organism and that the biogravitational field distorts the local subjective space-time of the conscious observer. . . . I conjecture that distortions can be manipulated in such a way that the rate of time

flow at the location of the participator does not match the corresponding rate of time flow at the object being observed and influenced. . . . The differential in the time flows of participator and object can in principle be so adjusted that the participator working within his local light cone "sees" into the probable future or past of the object (that is, he samples universe layers).[11]

and:

Telepathy can be understood as messages travelling through biogravitational wormholes of space-time within a given universe. The wormhole connections constitute only one possibility for a cosmic "telephone" network. In addition, the "feeling" or tachyonic model of gravitation that breaks through the wall of light also accounts for telepathic communication. In either case, the strength of telepathic communication should not be affected by distance.[12]

In spite of the difficulties of penetrating such terminology, however, it is worth the effort if only to appreciate that here we have a professor of theoretical physics saying the same thing, in rather different words, as the SF writer or the anecdotal reporter of visionary experiences and precognitive dreams. Our natural preference for the anecdotal material does not conceal any dubious basis in fact or a

[11] *Space-Time and Beyond.*
[12] Ibid.

lack of "respectable" scientific theorizing, but does indeed rest on a secure foundation.

In this connection, *Riddle of the Future* provides in two experiences of the wartime flier Sir Victor Goddard, almost perfect examples of what I regard as a form of mental communication sideways across the time barrier. One of these occurred in 1935 when, flying in an open-cockpit aircraft in rain and cloud, the then Wing Commander, Goddard, lost control and came within seconds of a fatal crash, recovering only just above ground level. Immediately afterwards, leveling out, he flew across the site of Drem airfield (close to Edinburgh), which was at that time (in "our" time track) a disused, rundown relic of the First World War. Suddenly seeming to be flying in bright sunlight, Goddard saw the airfield bustling with activity and aircraft lined up outside gleaming hangars. Equally suddenly, the "vision" vanished and he was back in driving rain and cloud.

Such a brief synopsis hardly does justice to the full story, which should be read in detail. But what I am interested in here is the explanation. Goddard himself came to regard the occurrence as a precognitive vision, because within a few years Drem was back in use as a training airfield as the RAF expanded to meet the needs of a Second World War—and used just the types of aircraft that Goddard had "seen" after his close encounter with death. But, many years later, he discovered that the hangars built when the airfield was reestablished were not of the older style that he had "seen," with bitumen-covered fabric roofs, but roofed with corrugated iron. Why did the vision go wrong?

Rather laboriously, Goddard suggested that his contact with the future had not been a direct vision of reality, but rather some form of image plucked from the mind of a *planner* (or planners) involved in the airfield's reconstruction, and that at some time it may have been intended to rebuild using the kind of hangars he "saw." This tortuous explanation develops entirely from the ingrained concept of one unique time track. Once we allow the possibility of parallel universes, things become much simpler. The vision now can be explained simply as a transfer sideways in time, to 1935 where, in summer sunshine, Drem airfield was still operational, or had been revitalized by a RAF rather more prepared for war than its counterpart in our Universe. Or perhaps Goddard did "see" the future, but a future "next door," in which different hangars were constructed. I prefer the first version—for where, after all, is the focus of human attention most likely to jump under threat of death in the here and now? Surely to the nearest equivalent universe where there is no threat of death, to the parallel world where the Sun shines and Goddard Number 2 flies serenely on his way?

The second paranormal experience relating to Goddard seems to tie in even more firmly with the existence of parallel worlds of alternative probability. This occurred in January 1946, when Goddard was in Shanghai and about to move on by air to Burma and Malaya. At a farewell cocktail party held in his honor, Goddard overheard a

Navy officer discussing with a friend a dream in which he (Goddard) had been killed in a crash. Naturally interested in this tale, Goddard questioned the Naval Commander about his dream; and although details remembered by the two men almost thirty years later don't tally in every respect, it is clear that, in front of many witnesses, the Navy officer reported waking from a dream with a firm conviction that Goddard had been killed in a crash, and being surprised to learn that this had not happened and that the party for Goddard was going ahead. Both officers had read J. W. Dunne's intriguing book *An Experiment with Time* (of which more in the next chapter) and recall remarking on his "rule of thumb" that precognitive dreams usually "come true" within two days or not at all, unless by coincidence.

A day later, the Dakota (DC-3) aircraft carrying Goddard and ten other people crashed on the shore of an island off the coast of Japan after a difficult flight in bad weather and cloud. *But none of the people on board were killed.* It is difficult to think of a better way to explain this kind of phenomenon than through "information" traveling in some way sideways across time. In our world, the plane did crash, and it would only have needed a slightly different pattern of probabilities for all on board to have been killed. Obviously, with such a small difference in probabilities involved, there must be not one but *many* nearby parallel worlds where the crash proved fatal. Indeed, *our* world may be the odd one out in this regard! Such a crash, in such circumstances, would (did!) make a marked impact on all the officers in Shanghai at the time; small wonder that, once again, the echoes should reverberate across and back through the barriers of time, to stir a response in one of those officers in our timestream.

Nothing, surely, can top that tale in this very brief survey of the anecdotal material. So it now seems appropriate to move once again into the realms of philosophy, and ponder on just how the focus of attention does shift across the timestreams—including a look at some of the most widely publicized ideas of the twentieth century on the nature of time, those of J. W. Dunne mentioned above.

CHAPTER
8
The Focus of Attention

In the middle decades of the twentieth century, probably the most widely read attempt to come to grips with the mysteries of time and the human mind was the work of J. W. Dunne. More recently, his writings have fallen into relative obscurity, not least because there is something not quite right about the theory underlying them. But they still stand as a remarkable body of modern work, giving us a mind-broadening insight into the mysteries we have been tackling throughout this book.

The starting point for Dunne's speculations about the nature of time was a series of dreams in which he seemed to be experiencing a dream mixture of past and future events. These were normal dreams, distorted compared with ''reality'' as all dreams are, but including aspects of events yet to be experienced, or "displaced in time." Many examples are given in Dunne's books,[1] but the overall tone of the experiences is very similar to those of other people recounted in Chapter Seven. What is particularly important in the context of a study of the links between time and the human mind, and the existence of timewarps, is that after a few such experiences Dunne began to make careful notes of his dreams on waking, and used these to check when later feeling any sense of déjà vu, of having "been here before" when visiting a strange place, and so on. He suggests that a mentally imposed barrier that operates only on the waking mind prevents us from "seeing" future events when awake, and he found that with training he could induce a kind of half-awake "daydream" state (rather like self-hypnosis) in which this innate ability could be trained to work, after a fashion, for the waking mind.

Dunne's greatest precognitive successes, however, remained those associated with dreams, both his own and those of others. His rare ability seems to have been not *having* precognitive dreams, but *remembering* both these and "ordinary" dreams, and keeping accurate notes about them. When friends were trained to observe with the same attention to detail, they too found dreams split almost exactly fifty-fifty between past and future events; and Dunne suggests that this is, in fact, the normal pattern in all of us—although many people don't even realize that they dream at all!

[1] See especially *An Experiment with Time*.

163

The theory he then elaborated does not depend on the dreams as such, but on the overall evidence for the existence of precognitive phenomena. The dreams provided the catalyst for his speculations about the nature of time, which build from the suggestion, often pondered on by philosophers, that the idea of "movement" through time from the past toward the future must imply the existence of some second-level "supertime" that is used to measure the rate at which ordinary time passes. Then, of course, you need a "super-supertime" to measure the flow of supertime, and so on ad infinitum. This infinite series of different "times" (T_1, T_2, T_3 and so on) gives Dunne's theory its name "serial time," which looks quite neat except for the puzzle of what really happens "at infinity." Another way of looking at the puzzle is to extend the concept of time as the fourth dimension, at right angles to each of the three dimensions of space we know so well. This would be T_1; T_2, by analogy, must "be" at right angles to all four of the other dimensions, in the fifth dimension; T_3 corresponds to the sixth dimension, and so on again ad infinitum.

Now, on this picture an observer who exists in *four* dimensions would see "our" time, T_1, as part of space, with future and past visible like forwards or backwards on a road, or up and down in space. A five-dimensional observer would "see" everything in the series below T_3, and so on. The layers of time are, in a supradimensional sense, nested one inside the other rather than running alongside as parallel universes; and the ultimate "real" observer of everything is a mind or consciousness with access to the infinite variety of times, a supermind whose attention focuses briefly on our humdrum four-dimensional world to give us the illusion of existing in three dimensions of space and one of time.

The self of our everyday world is, if you like, a three-dimensional cross-section of a greater, multifaceted self—although there remains the question of whether we each belong to a separate supermind, or whether, coming close to Hoyle's more recent speculations, there is one super-mind whose attention flickers across all our lives, making us all different three-dimensional facets of the *same* ultimate consciousness. Why should the "ultimate" observer focus attention on simple little T_1? Because, says Dunne, this *is* the simplest; the infant superconsciousness starts to learn here, but does not end here since "we must die before we can hope to advance to a broader understanding."[2] Here is the implicit recognition of an immortal soul required by the seriality of time in Dunne's model. We start out in T_1 and move onward and upward as the soul "grows" into maturity.

Apart from anything else, this rather simple picture certainly runs into difficulties at a very mundane level when we consider the "birth," growth, and death of, say, a tree—do trees have multidimensional souls?—and at a high level because Dunne seems to think of infinity as a definite place where the supermind or superminds can be located. But the echo of hierarchical religions is intriguing, as is the possible relation to reincarnation—if at "death"

[2] *The New Immortality.*

we haven't learned enough to move up a notch in the series, do we have to go back for another practice run in T_1?

Although Dunne produced several more books elaborating his theme, he never succeeded in producing a version of serial time that was really satisfactory, largely because of the difficulty of explaining what is really meant by "infinity." Some of the problems are discussed by MacKenzie.[3] But what Dunne did achieve was enormous publicity for the existence of precognitive phenomena, which encouraged many people to keep records of dreams and directly resulted in information about many precognitive dreams becoming available—as we saw in the previous chapter, an unusual dream experienced in the 1940s by a very solid member of the community (a serving officer in the Navy) was discussed and remembered by others in large part because they knew of Dunne's ideas.

One idea from a later book,[4] however, is notable because it brings us very close again to the idea of one supermind whose attention flickers across an array of pigeonholes representing our lives or events in our lives. This idea brings in a particularly happy analogy for a writer, that of the keyboard on a standard typewriter. The letters of the keyboard are capable of providing an almost infinite array of information, but in themselves they don't tell us much. If we looked on this series as representing a flow of "information," leaving aside the numbered keys, the pattern we have runs from "QWERTY . . ." at the top to

[3] *Riddle of the Future.*
[4] *The New Immortality.*

". . . CVBNM" at the bottom, and doesn't make a lot of sense. But if the "focus of attention" shifts from key to key not in a linear fashion but obeying some greater law (corresponding to a higher dimension in which attention can move backwards or forwards across the keys), then the result is meaningful prose. Not just that, but the same keyboard can be used time and time again to provide different sets of meaningful prose—or, indeed, different sets of gibberish. It is the focus of attention that is all-important—and the same may well be true of our perception of the Universe and of the "flow" of time.

Jung and the Collective Unconscious

Maybe you're a bit uneasy about the credentials of Dunne, an aeronautical engineer, or even Hoyle, an astrophysicist, to offer theories about the workings of the human mind and the nature of our perception of reality. But if you want support for this broad picture of consciousness as some aspect of a greater "mind," you don't have to look far among the ranks of the most eminent authorities on the human mind to find it in the works of Carl Gustav Jung, referred to in passing at the beginning of Chapter Seven. As well as his own experiences of meaningful acausal "coincidences" or "synchronicities," Jung has discussed the precognitive dream experiences of Dunne, the classic experiments in extrasensory perception and other psi phenomena (such as those of Rhine),[5]

[5] See *Synchronicity*.

and—perhaps of greatest interest in the present context—
the links with traditional "Eastern" patterns of thought
and philosophy, as typified by the body of work that has
come down to us as the *I Ching*.[6]

Jung's key contribution to mainstream psychology is, of
course, his modification of the earlier ideas of Freud to en-
compass the concept of the "collective unconscious."
Where Freud made a major breakthrough in understanding
the nature of the human mind by realizing that each of
us has his own personal unconscious mind that affects his
conscious thought and reactions to different circumstances,
Jung proposed that we all *share* a collective unconscious
built up from all of the memories and patterns of behavior
that have grown up over the entire history of mankind. If
we all share the same collective unconscious, then it is no
surprise that we are all predisposed to respond in the same
way to certain stimuli. In general, people are afraid of the
dark and of snakes, even though they may live in a modern
city with electric lighting and no snakes to be seen. The
Jungian explanation of this phenomenon is that the night-
time and snakes were major dangers to primitive man, and
that these once relevant fears have been passed on through
the collective unconscious.

This concept has mystical or religious overtones, and it
is hardly surprising that Jung's work has been a major influence in discussions of philosophy and religion. Given
the view of the nature of time that is being built up in the
present book, we can also see that perhaps the collective

[6] See Jung's Foreword to the Richard Wilhelm translation of the *I Ching*.

unconscious is not something that is merely handed on from generation to generation, but a wide-ranging (in time) pool of information that is tapped by every human mind, the "tuning in" in a general and undefined way to everything that has gone before. Cases of reincarnation, or precognition, may then represent rarer *precise* resonances between individual minds separated in time but swapping very detailed information about specific events. We see also that in all probability the collective unconscious should be viewed as a collection not only of *past* experiences but of *future* experiences as well, with the fears and hopes of generations yet to come also gathered in the whole ebb and flow of the unconscious mind. To extend the example of snakes and night bringing echoes of fear from the distant past, is it possible that the human obsession with extraterrestrial travel and contact with other intelligent beings echoes, backwards down the corridors of time, the day when we will indeed make such contact, producing enormous repercussions on human society and a dramatic impact on the collective unconscious? A similar innate awareness of the possible growth of probability paths from the present state gives a new insight into the value of the *I Ching* as a tool for understanding the workings of the human mind, while tying the Eastern view of reality firmly in with modern Western ideas about the nature of time derived from our "scientific" approach.

The *I Ching* and the Eastern Way

The difference between the two approaches in philosophical terms is essentially that where the Western way is to take a collection of details and use them to build up a broad picture, the Eastern way is to attempt to grasp the broad view, against which details can be seen in their wider context. The *I Ching—Book of Changes—*is an aid to this process of gaining an insight into the broad view of the whole situation.

We have already looked briefly at the Eastern philosophical basis of the Tao. Jung puts this in perspective in a succinct summing up:

> In Chinese philosophy one of the oldest and most central ideas is that of Tao, which the Jesuits translated as "God." But that is correct only for the Western way of thinking. Other translations, such as "providence" and the like, are mere makeshifts. Richard Wilhelm brilliantly interprets it as "meaning." The concept of Tao pervades the whole philosophical thought of China. Causality occupies this paramount position with us, but it acquired its importance only in the course of the last two centuries, thanks to the leveling influence of the statistical method on the one hand and the unparalleled success of the natural sciences on the other which brought the metaphysical view of the world into disrepute.[7]

[7] *Synchronicity.*

He goes on to quote a particularly apposite section from the *Lao Tzu,* the classic work on Taoist thought, in which Lao Tzu discusses the significance of the "emptiness" or "nothing" that is essential to the whole and is at the heart of the Tao:

> We put thirty spokes together and call it a wheel;
> But it is on the spaces where there is nothing that
>   the utility of the wheel depends.
> We turn clay to make a vessel;
> But it is on the space where there is nothing that
>   the utility of the vessel depends.
> We pierce doors and windows to make a house;
> And it is on these spaces where there is nothing that
>   the utility of the house depends.
> Therefore just as we take advantage of what is, we should
>   recognize
>   the utility of what is not.[8]

Traditionally, this kind of concept has been discussed and applied within a framework of space and material things—even Capra in his *Tao of Physics* is concerned primarily with "things," although just what the reality of things is becomes an interesting question. But, again, we can extend the idea to our concept of parallel worlds of alternative probability, branching away from the here and now, the focus of our attention, in their infinite variety. If we try to find some guidance to the best or most desirable future path in a particular situation, we cannot look at the

---

[8] Quoted from *Synchronicity.*

future that *is,* only at the futures that *may be,* even granted the ability of the (collective?) unconscious to warp time. If we can obtain information about possible futures, what "is not" becomes of crucial importance, since our actions, resulting from awareness of future possibilities, will inevitably direct the focus of our attention along one path at the expense, from our point of view, of many other future worlds that fail to become real or "cease to exist." The fact that our alter egos, the focus of the supermind that passes along these alternative futures, see a different concept of reality in which "our" future is one of the many that fail to achieve existence, makes no difference from the point of view of a consciousness with knowledge essentially restricted to one developing time track. From whichever track the situation is viewed, the "what is" and the "what is not" are of equal importance.

Does this help us in understanding the real nature of the *I Ching,* the book which has gained a crude popularity in the West recently as a kind of fortune-telling device? The answer, most definitely, is yes. There is far more to the *I Ching* than crude fortune telling, although the fortune-telling aspect of the *I Ching* is echoed by Western concepts such as the Tarot that date from the period before the rise to prominence of the central concept of causality referred to by Jung in the quotation above.

This is not the place for a detailed description of the *I Ching,* something which I am not, in any case, competent to give. A good basic guide to the simplest way of gaining access to the wisdom embodied in this work is provided by Alfred Douglas in my personal favorite, *How to Consult the I Ching,* and by Da Liu in *I Ching Coin Prediction;* and there are many other similar volumes available. Put very simply, however, the *I Ching* contains sixty-four "hexagrams," each a symbol composed of six lines, either broken or unbroken (see Figure 6.6). These sixty-four symbols are each "read" as a pair of three-line symbols or trigrams, and each symbol has its own commentary handed down from the philosophers of ancient times. In passing, it is interesting to realize that this is at heart a basic "binary" code very similar to the way information is coded for processing by modern high-speed computers. In modern electronic computers, the basis is a "yes/no" code corresponding to whether each of an array of switches in a network is on or off; in the *I Ching* trigrams, the code depends upon which lines are broken or unbroken, exactly equivalent to the simple binary yes/no choice. There is no deep significance to this, however; it doesn't mean the ancient Chinese had electronic computers or were visited by spacemen who used the binary code. It's just that this is the most basic method of encoding information, so fundamental that it was hit upon independently by civilizations as diverse as those of ancient China and our modern scientific culture!

For simple fortune telling—an abuse of the *I Ching,* but the initial introduction to its philosophy experienced by most people today—the approach is to "ask" the book about the outcome of a proposed course of action, framing the question either mentally or out loud, while tossing a set of three coins (or using some other system, but the coin method is simplest) to select on the basis of the fall of heads and tails which hexagram should be referred to. The commentary corresponding to that hexagram will give an insight into the outcome of the proposed course of action—but it will not "tell" you what to do. Rather it expresses in deeper terms the relation of your problem to the broader reality, and draws attention, perhaps, to the nothing or "what is not" that surrounds and interacts with the distracting detail of your specific question. Viewed from this new perspective, the most difficult problems can become easily resolved. But to what extent is this "fortune telling" or "precognition," and how far can the results simply be interpreted as the natural consequence of pausing thoughtfully to appraise the difficult situation from a different point of view?

Well, you pay your money and you take your choice. Or do you? If there really is a way for the mind, or supermind, to tune in to alternative future possibilities and probabilities, then the kind of conditions involved in such an exercise—the darkened room, the concentration on the question at hand, the mystical rhythm of the coin-tossing ritual—are just the conditions appropriate for tuning in to whatever it is we tune in to. Done properly, as devotees insist it should be done, the ritual is well on the road to self-hypnotism, and certainly concentrates the mind wonderfully.

Perhaps it is appropriate to give an example. I have just asked the oracle the following question:

Am I wise in attempting to relate Eastern philosophy and Western science in this study of timewarps?

Not an easy question, or particularly well phrased, but one that reflects my own major concern at this moment, and should therefore be ideal material for consideration. The "answer" I received is hexagram number 20, Kuan (Looking down)

with indications that this should be taken in conjunction with hexagram 61, Chung Fu (Inward Confidence and Sincerity).[9]

The second, especially, looks encouraging. But what are the detailed commentaries on these hexagrams? The commentaries that follow are taken from John Blofeld's *I Ching: The Chinese Book of Change,* with some of the subtleties of the significance of specific lines omitted. For hexagram 20 we find the commentary:

> Looking down in its most important sense means that looking down which takes place from on high. Willing acceptance and mildness are conjoined . . . the text refers to those who look down on their subjects and transform them. They contemplate the sacred ac-

[9] The way a hexagram gives rise to a secondary hexagram is a result of the details of the fall of the coins, explained in all the standard *I Ching* volumes.

tivities of heaven and note how the seasons unfold, each in its proper time. It is because the holy sage makes these matters the subject of his teaching that all the world accepts his dominion.

The oracle, it seems, has an almost embarrassing regard for the importance of my stumbling attempts to interpret for others the mystery of time;[10] and the appropriateness of the commentary to the question asked and to my own position as an "academic" (the modern equivalent of a holy sage) who has now turned to authorship is astonishing—to anyone not used to the bull's-eye accuracy of the *I Ching*.

But what of the related hexagram, number 61? Here we find indications of:

> Inward confidence and sincerity. Dolphins—good fortune! It is advantageous to cross the great river (or sea). Persistence in a right course brings reward.

Indications, perhaps, of a greater response to this book in the United States than in Britain? And the further commentary runs:

> Joyfulness and gentleness are conjoined and there is sufficient confidence to ensure the smooth develop-

---

[10] The Alfred Douglas translation is even more positive: "The superior man contemplates the law of the universe and the recurring cycle of the seasons. Acting as a vehicle for the power of Heaven, he achieves greatness." I happened to have the Blofeld volume at hand when I asked my question, and only later looked up the appropriate commentary in Douglas—but now you know why Douglas is my favorite!

ment of the realm. The good fortune symbolised by dolphins is that of winning the confidence of every creature. . . . Persistence is always advantageous when it is accompanied by confidence, for then it accords with heaven.

All this could hardly be more encouraging, especially at a time when the book is nearly complete and I am slightly worried by the usual doubts an author has in the interval between preparing a work and its publication.[11] I am assured that I am doing the right thing, that I must remain confident and persistently put forward my ideas, and that like other "wise men" I am of value to society even though I do not produce any material benefits. In such terms, this is the kind of advice and commentary that might be expected from a psychologist who knew me and my work well; yet these are the translated words of a long-dead Chinese sage, plucked from a wide choice in a particular book as a result of tossing three coins six times! It is hardly that all of the commentaries say the same kind of bland, reassuring things—take number 12, P'I (Stagnation), for example:

Stagnation (or obstruction) caused by evil-doers. . . . The great and the good decline; the mean approach . . . the celestial and terrestrial forces are without intercourse and . . . everything is out of communion with everything else.

[11] The Douglas version, a little more succinct, includes in the commentary, "In the virtuous exercise of firm correctness we see the proper response of man to the influence of Heaven."

That would have been an equally reasonable response to my question, if my work had so confused the issue as to be unworthy of further discussion. The fact is, though, that without giving a simple yes or no answer the oracle provided me with encouragement and astute guidance to the best path to follow in future. My own belief is that this "works" because in asking the question my subconscious was stimulated to make contact with the collective unconscious (or the supermind, or whatever it is that gives us access to information about possible future worlds), and through some feedback we do not yet understand in terms of physical science my mind (or the supermind) then influenced the fall of the coins to guide my conscious self toward the appropriate *I Ching* commentaries.

Stated that baldly, it sounds crazy to anyone trained in modern Western physical science, at least at a basic level. But think again on the implications of relativity and quantum theory, on the bizarre frontier lands now being explored by conventional physical science and touched on in Part Two of this book. If the focus of attention is all-important, as I believe it is in some transcendental "cosmic" sense, then why shouldn't we view the mechanism of coin tossing, or of the Tarot pack, as an aid to concentration, to direct the focus of attention along an appropriate line? If Jung was prepared to put his head on the chopping block for such nonmainstream views, then surely so should I, echoing his words:

> It is a dubious task indeed to try to introduce a collection of archaic "magic spells" to a critical modern

public. . . . I have undertaken it because I think there is more to the ancient Chinese way of thinking than meets the eye. But it is embarrassing to me that I must appeal to the good will and imagination of the reader, instead of giving him conclusive proofs and scientifically watertight explanations.[12]

And, equally apposite in the present context:

The Chinese mind, as I see it at work in the *I Ching,* seems to be exclusively preoccupied with the chance aspect of events. What we call coincidence seems to be the chief concern of this peculiar mind, and what we worship as causality passes almost unnoticed.[13]

The chance aspect of events—alternative worlds of probability—quantum mechanics—statistical probability. All are inextricably intertwined, and it is to Eastern philosophy that we must turn in an attempt to come to grips with what our "modern" scientific approach is now beginning to tell us. Our pursuit of the twin questions "What is time?" and "Can we travel in time?" has not, perhaps, led entirely in expected directions. But it has certainly led us in important and significant directions, giving us new insights into the whole nature of reality and the place of the scientific approach in a broader context. All we are left

[12] "Forward to the 'I Ching' " in *Psychology and Religion.*
[13] Ibid.

with—for now—is to take stock of the path we have followed, and see just what kind of brief, snappy answer we can provide to our original question—while appreciating, perhaps, that answers in the sense that we began to look for them are no longer either appropriate or relevant.

CHAPTER
9
Mind over Time

So we come back to the basic question "Is time travel possible?" In the light of the evidence gathered together in this book, my answer would be, "Yes, it happens all the time." But to most people such an answer would mean something different from the meaning I put on it, because most people, if they think about time travel at all, think about the Wellsian image of a machine for moving through time—or, at a slightly more sophisticated level, of the mysteries of black holes and relativity theory discussed in Part Two. As I mentioned in the Introduction (but can't resist reiterating, not least since many people don't read the introductions to books!) the approach to the mysteries of timewarps that I have used here closely follows the development of my own interest in such phenomena, over a period of many years. Looking beyond the commonsense world of time flowing like a river, we find the way-out world of the modern researchers, where time behaves in most peculiar fashion. But as this area of study shades into puzzles beyond (and more fundamental than) the physical sciences, the question of our whole perception of reality becomes part and parcel of the mystery of the nature of time. Questions such as the nature of consciousness and the possible existence of a supermind simply are not tackled by the physical sciences, although that doesn't stop individual scientists, such as Hoyle, from looking beyond the narrow confines of their own disciplines to the broader cosmic horizons.

Time "travel" with or within the mind is certainly the best immediate prospect. The simple training and self-discipline needed to make constructive use of the information about both past and future that is fed into our awareness by the subconscious through the dream mechanism is within the reach of anybody who has stuck with me this far and who is tempted to read further, perhaps in Dunne's books. But can this mental ability (or perception, depending on how you look at it) be extended into something closer to what we think of conventionally as "time travel"? Once again, continuing the theme that has proved so useful to us throughout this investigation of the mysteries of time, we can find some tantalizing prospects by looking at the imaginations of some science fiction writers and setting them in the perspective of philosophies about cosmic awareness, parallel universes, and the possibility of a supermind or -minds.

182

Clifford Simak, in his novel *Time Is the Simplest Thing,* neatly highlights the relationship between space and time, as well as dealing with the more mundane SF fare of a maverick hero, fighting against a corrupt society, who saves the world (or at least the nice people in it) through the discovery of latent superpowers in his own mind. Take that as you will—at least it makes a good read—but don't throw out the "philosophy" unthinkingly. Simak's "hero" is involved with the successor of modern space projects, the development that followed with the realization that there was no way to crack the speed-of-light barrier and that mankind was chained physically forever within the confines of one small solar system. Sending out machines instead—impervious to radiation and for all practical purposes immortal—mankind maintains exploration of the Universe through mental contact (aided by other machines) with the robot explorers. In the process, the hero makes direct mind-to-mind contact with an intelligent alien, learns how to do all of this and more without the aid of machines, and the plot is off and running.

But the point is a serious one. Time travel is as "impossible," according to the standard picture, as practicable space travel between the stars. And if you can provide a means of transferring the focus of attention across space, then you are automatically involved in providing a means of shifting the focus of attention through time, up or down that stereotyped river or sideways into the river next door. Turning the thing on its head, remember that if we think of time as the "fourth dimension," then in a sense the "length" of each second, in terms of that universal constant the speed of light, is 186,000 miles. To be sure, physical time travel may be possible—but if you want to get anywhere you're going to have to "travel" at quite a "speed" to do it. Unless, that is, you can nip round the back of a convenient black hole, or find a way of juggling with tachyons to produce faster-than-light travel.

Roger Zelazny, in his five "Amber" stories,[1] goes whole hog in avoiding such complications by giving his characters—some of them—the ability to cross into parallel universes at will through the power of mind alone, or in the terms I have used in this book, through a conscious shift in the focus of attention. He doesn't get them involved in any subtleties of forward and backward travel through time, the sideways form being sufficient to muddy up his plot. And the process isn't presented as moving through parallel worlds but as bringing into reality different shadows of the central, dominant world line. (As in H. Beam Piper's tales and others, "our" Universe is merely one of the secondary shadows.)

> Something told me that whatever Shadows were, we moved among them even now. How? It was something Random was doing, and since he seemed at rest physically, his hands in plain sight, I decided it was something he did with his mind. Again, how?
>
> Well, I'd heard him speak of "adding" and "subtracting," as though the universe in which he moved were a big equation. I decided—with a sudden certainty—that he was somehow adding and subtracting

---

[1] *Nine Princes in Amber, The Guns of Avalon, The Sign of the Unicorn, The Hand of Oberon,* and *The Courts of Chaos.*

items to and from the world that was visible about us
to bring us into closer and closer alignment with that
strange place Amber, for which he was solving.[2]

Nice work if you can get it, or if you were born a superbe-
ing with such exotic powers. Changing a world—or universe—by the power of mind seems a bit hard to swallow.
But it looks more reasonable from the point of view of
shifting one's awareness into the next-door pigeonhole, or
the adjacent stack of pigeonholes.

Alan Garner, in his compelling book *Red Shift*, looks at
the bit of the puzzle Zelazny leaves out—travel up and
down one river of time (or, at least, communication across
the time barrier) through the power of mind, with no extra
complications from parallel universes. In this story the
central character is linked—through a focus of attention on
a strange, significant stone axe from the ancient past—to
the lives of a Stone Age man, a character living in the af-
termath of Roman Britain, and a young man in medieval
England, all of them experiencing violent lives linked to
the power of the Beaker axe. This, of course, closely
echoes the real life "reincarnation" investigations such as
those discussed in Chapter Seven. The thread of recurring
violence is singling out lives from the many available to be
scanned by the unconscious mind, with the stone axe in
this story as an added aid to concentrating the unconscious
focus of attention, and providing a running thread linking
past with present—and future?

[2] *Nine Princes in Amber.*

We are still a long way from direct, conscious control of such phenomena—but a lot closer, perhaps, than we are to building a physical time machine. The subjectivity of time is clear to anyone who has dealings with children, or remembers his own childhood; to a child, a year is an enormous span of time, and even planning for something a week or two ahead is comparable, in our adult world, to laying down a time capsule for posterity. Even in the adult world—and including the world of science—things do not follow the simple pattern we like to pretend exists. Scientists are familiar with the sudden insight into the nature of a problem that comes without all the logical, ordered steps in the puzzle having been solved—the flash of intuition that enables a person to go back and fill in the logical details afterwards, dressing them up for publication in learned journals along the lines of "If *this* happens then *this* happens and *that* results." L. R. B. Elton and H. Messel discuss the phenomenon in their book *Time and Man:*

> The conscious thought processes in mathematics are unidirectional not only in time but also in logic, and are therefore particularly suitable for study from our point of view. Now, again and again, mathematicians have recorded their experience of the sudden flash of insight. This commonly comes quite unexpectedly, after a long process of conscious thought on a problem had not led to its solution. The flash of insight then suddenly presents the whole solution in a way quite unconnected with the previous conscious thought processes, and all at once, rather than over a

time interval in logical order. The experience is of course not confined to mathematicians, but forms a crucial part in the act of creation in any scientific activity and in many non-scientific activities. One of the authors, for instance, has experienced it in connection with the translation of poetry. Its most important characteristic is the simultaneity with which the complex solution arrives in the conscious mind, which is an indication of the timeless nature of the unconscious.

For someone trained in physics and astronomy, this is the most important realization to emerge from an attempt to come to grips with the nature of time. There is indeed far more to the Universe (or universes) than our physical senses and physical sciences can reveal, and what we see as reality may indeed be largely subjective, conditioned by our preconceived ideas and those of the society in which we live. Timewarps exist. *Control* of timewarps (and spacewarps), however, is something to be sought not with the aid of mechanical devices, but through the improved understanding of the human mind, the nature of the unconscious, and their interactions with what we think of as the physical world. And the key to this progress probably lies in Eastern ''philosophy'' rather than Western ''science,'' although, as Capra has commented in *The Tao of Physics,* there is less difference between the two than meets the eye.

187

BIBLIOGRAPHY

ALEXANDER, S. *Space, Time and Deity.* London: Macmillan, 1927.

AMIS, KINGSLEY. *The Alteration.* New York: Viking, 1976.

ATKINS, KENNETH R. *Physics: Once Over Lightly.* New York: Wiley, 1972.

BERGSON, HENRI. *Time and Free Will.* Seventh impression. London: Allen and Unwin, 1959.

BERRY, ADRIAN. *The Iron Sun: Crossing the Universe through Black Holes.* London: Cape, 1977.

BLOFELD, JOHN (translator). *I Ching: The Chinese Book of Change.* New York: E. P. Dutton, 1974.

BUTTLAR, JOHANNES VON. *Journey into Infinity.* London: Fontana, 1976.

BIBLIOGRAPHY

CAPRA, FRITJOF. *The Tao of Physics*. New York: Bantam, 1977.

CARROLL, LEWIS. *Through the Looking Glass*. In *The Annotated Alice* edited by Martin Gardner. New York: New American Library, 1974.

CLEUGH, MARY F. *Time and Its Importance in Modern Thought*. London: Methuen, 1937.

CLUTTON-BROCK, M. "Entropy per Baryon in a 'Many-Worlds' Cosmology." *Astrophysics and Space Science* 47 (1977): 423.

DA LIU. *I Ching Coin Prediction*. London: Routledge and Kegan Paul, 1975.

DAVIES, PAUL. "Gödel and General Relativity." *New Scientist*, 26 January 1978: 239.

DAVIES, PAUL. *The Runaway Universe*. London: Dent, 1978.

DAVIES, PAUL. *Space and Time in the Modern Universe*. Cambridge: Cambridge University Press, 1977.

DE CAMP, L. SPRAGUE. *Lest Darkness Fall*. New York: Ballantine, 1974.

DICK, PHILIP K. *Counter-Clock World*. New York: Berkley Medallion, 1967.

DICK, PHILIP K. *The Man in the High Castle*. New York: Berkley Medallion, 1974.

DOOB, LEONARD W. *Patterning of Time*. New Haven: Yale University Press, 1971.

DOUGLAS, ALFRED. *How to Consult the I Ching*. London: Penguin, 1972.

DUNNE, J. W. *An Experiment with Time*. Atlantic Highlands, N.J.: Humanities Press, 1959.

DUNNE, J. W. *Intrusions*. London: Faber and Faber, 1955.

DUNNE, J. W. *The New Immortality*. London: Faber and Faber, 1938.

DUNNE, J. W. *Nothing Dies*. London: Faber and Faber, 1940.

ELTON, L. R. B., and H. MESSEL. *Time and Man*. Oxford: Pergamon Press, 1978.

GALE, R. M. *The Language of Time*. Atlantic Highlands, N.J.: Humanities Press, 1968.

GARNER, ALAN. *Red Shift*. New York: Macmillan, 1973.

GAUQUELIN, MICHEL. *The Cosmic Clocks: From Astrology to a Modern Science*. Chicago: Contemporary Books, 1974.

GERROLD, DAVID. *The Man Who Folded Himself*. Mattituck, N.Y.: Aeonian Press, 1976.

GRIBBIN, JOHN. *What's Wrong with Our Weather?* New York: Scribner's, 1978.

GRIBBIN, JOHN. *White Holes: Cosmic Gushers in the Universe*. New York: Delacorte, 1977.

HALDEMAN, JOE. "End Game." *Analog* XCLV, No. 5 (January 1978): 66.

HALDEMAN, JOE. *The Forever War*. New York: Ballantine Books, 1976.

HARRISON, HARRY. "Worlds Beside Worlds." In *Explorations of the Marvellous* edited by Peter Nicholls. London: Fontana, 1978.

HAWKINS, GERALD S. *Beyond Stonehenge*. New York: Harper and Row, 1973.

HAWKINS, GERALD S. *Stonehenge Decoded*. New York: Dell, 1966.

HEINLEIN, ROBERT. "All You Zombies." In *The Best of Robert Heinlein 1947–59*. London: Sphere, 1973.

HEINLEIN, ROBERT. "By His Bootstraps." In *The Astounding-Analog Reader, Book One* edited by Harry Harrison and Brian W. Aldiss. London: Sphere, 1973.

HILGARD, E. R., R. C. ATKINSON and R. L. ATKINSON. *Introduction to Psychology*. Sixth Edition. New York: Harcourt Brace Jovanovich, 1975.

HINCKFUSS, IAN. *The Existence of Space and Time*. Oxford: Clarendon Press, 1975.

HODGSON, S. W. *Time and Space*. London: Longman, 1865.

HOYLE, FRED. *From Stonehenge to Modern Cosmology*. San Francisco: W. H. Freeman, 1972.

HOYLE, FRED. *October the First Is Too Late*. New York: Harper and Row, 1966.

HOYLE, FRED. *Ten Faces of the Universe*. San Francisco: W. H. Freeman, 1976.

HOYLE, TREVOR. *The Gods Look Down*. London: Panther, 1978.

HOYLE, TREVOR. *Seeking the Mythical Future*. London: Panther, 1977.

HOYLE, TREVOR. *Through the Eye of Time*. London: Panther, 1977.

IVERSON, JEFFREY. *More Lives Than One?* New York: Warner Books, 1977.

JUNG, C. G. "Foreword to the 'I Ching.' " In *Psychology and Religion: East and West. The Collected Works of C. G. Jung,* translated by R. F. C. Hull. Princeton: Princeton University Press, 1970.

JUNG, C. G. *Synchronicity: An Acausal Connecting Principle,* translated by R. F. C. Hull. Princeton: Princeton University Press, 1973.

KAUFMANN, WILLIAM J. *The Cosmic Frontiers of General Relativity.* Boston: Little, Brown and Company, 1977.

KOESTLER, ARTHUR. *The Roots of Coincidence.* New York: Random House, 1972.

LAKE, K. and R. C. ROEDER. "Blue-Shift Surfaces and the Stability of White Holes." *Lettere al Nuovo Cimento* 16, No. 1 (1976): 17–21.

LAKE, K. and R. C. ROEDER. "Effects of a nonvanishing cosmological constant on the spherically symmetric vacuum manifold." *Physical Review Digest* 15, No. 12 (1977): 3513–3519.

LAKE, K. and R. C. ROEDER. "White Holes of Types II and III." *L'evolution des galaxies et ses implications cosmologiques.* Proceedings of CNRS International Colloquium No. 263. Paris: Editions du Centre National de la Recherche Scientifique, 1977.

LAKE, KAYLL. "White Holes." *Nature* 272 (1978): 599.

LAUMER, KEITH. *Dinosaur Beach.* New York: Scribner's, 1971.

LAUMER, KEITH. *The World Shuffler.* New York: Berkley Medallion, 1970.

MACKENZIE, ANDREW. *Riddle of the Future: A Modern Study of Precognition.* New York: Taplinger, 1975.

MARDER, L. *Time and the Space-Traveller.* Philadelphia: University of Pennsylvania Press, 1974.

MARTIN, GEORGE R. R. "Nor the Many-Colored Fires of

a Star Ring.'' In *Faster than Light* edited by Jack Dann and Georg Zabrowski. New York: Harper and Row, 1976.

MENZEL, DONALD J. *Astronomy*. New York: Random House, 1970.

MILES, V. W., H. O. HOOPER, E. J. KACZOV, and W. H. PARSONS. *College Physical Science*. Third edition. New York: Harper and Row, 1974.

MISNER, C. W., K. S. THORNE and J. A. WHEELER. *Gravitation*. San Francisco: W. H. Freeman, 1973.

MOORCOCK, MICHAEL. *Behold the Man*. New York: Avon, 1976.

MOORCOCK, MICHAEL. *The Time Dweller*. London: Mayflower, 1971.

MOORE, PATRICK (editor). *1976 Yearbook of Astronomy*. New York: Norton, 1976.

MOORE, WARD. *Bring the Jubilee*. New York: Avon, 1976.

NARKILAR, JAYANT. *The Structure of the Universe*. Oxford: Oxford University Press, 1977.

NEEDHAM, JOSEPH and WANG LING. *Science and Civilisation in China, Volume Three: Mathematics and the Sciences of the Heavens and the Earth*. Cambridge: Cambridge University Press, 1959.

NEEDHAM, JOSEPH, WANG LING and DEREK J. DE SOLLA PRICE. *Heavenly Clockwork*. Cambridge: Cambridge University Press, 1960.

NICHOLLS, PETER (editor). *Explorations of the Marvellous*. London: Fontana, 1978.

NIVEN, LARRY. "The Borderland of Sol." *Analog* XCLV, No. 5 (January 1975): 12.

NIVEN, LARRY. *A World Out of Time*. New York: Holt, Rinehart and Winston, 1976.

PALMER, JOHN D. "Biological Clocks of the Tidal Zone." *Scientific American* 232, No. 2 (February 1975): 70.

PAUWELS, LOUIS and JACQUES BERGIER. *The Morning of the Magicians*. New York: Stein and Day, 1964.

PRIEST, CHRISTOPHER. *The Space Machine*. New York: Harper and Row, 1976.

RENFREW, COLIN. *Before Civilisation: The radiocarbon revolution and prehistoric Europe*. London: Jonathan Cape, 1973.

ROBERTS, KEITH. *Pavane*. New York: Berkley, 1977.

RONAN, COLIN. *The Ages of Science*. London: Harrap, 1966.

SANTILLANA, GIORGIO DE and HERTHA VON DECHEND. *Hamlet's Mill*. Boston: Gambit, 1969.

SILVERBERG, ROBERT. *Up the Line*. London: Sphere, 1975.

SIMAK, CLIFFORD D. *Time and Again*. New York: Ace, 1976.

SIMAK, CLIFFORD D. *Time Is the Simplest Thing*. London: Victor Gollancz, 1962.

SMYLIE, D. E. and L. MANSINHA. "The Rotation of the Earth." *Scientific American* 225 (December 1971): 80.

TAYLOR, EDWIN F. and JOHN A. WHEELER. *Spacetime Physics*. San Francisco: W. H. Freeman, 1966.

TAYLOR, JOHN. *Black Holes*. London: Souvenir, 1973.

TOBEN, BOB. *Space-Time and Beyond*. New York: Dutton, 1975.

TROTSKY, LEON. *My Life*. New York: Pathfinder Press, 1970.

ULLMAN, MONTAGUE, STANLEY KRIPPNER and ALAN VAUGHAN (editor). *Dream Telepathy: Experiments in Nocturnal ESP*. London: Penguin, 1974.

WATSON, LYALL. *Gifts of Unknown Things*. New York: Simon and Schuster, 1977.

WATSON, LYALL. *Supernature*. New York: Bantam, 1974.

WEINBERG, STEVEN. *The First Three Minutes: A Modern View of the Origin of the Universe*. New York: Basic Books, 1976.

WHITE, JAMES. *Monsters and Medics*. New York: Ballantine, 1977.

WHITROW, GERALD. *The Nature of Time*. New York: Holt, Rinehart and Winston, 1973.

WILLIAMSON, JACK. *The Legion of Time*. London: Sphere, 1977.

ZELAZNY, ROGER. *The Courts of Chaos*. New York: Doubleday, 1979.

ZELAZNY, ROGER. *The Guns of Avalon*. New York: Avon, 1974.

ZELAZNY, ROGER. *The Hand of Oberon*. New York: Doubleday, 1976.

ZELAZNY, ROGER. *Nine Princes in Amber*. New York: Avon, 1977.

ZELAZNY, ROGER. *Sign of the Unicorn*. New York: Doubleday, 1975.

INDEX